# INDUSTRIALIZATION
### and the
## AMERICAN LABOR MOVEMENT
## 1850-1900

Kennikat Press
**National University Publications**
Series in American Studies


*General Editor*
James P. Shenton
*Professor of History, Columbia University*

IRWIN YELLOWITZ

# INDUSTRIALIZATION

## and the
## AMERICAN LABOR MOVEMENT
## 1850–1900

National University Publications
KENNIKAT PRESS      //      1977
Port Washington, N. Y.      //      London

Manufactured in the United States of America

Published by
Kennikat Press Corp.
Port Washington, N. Y./London

Library of Congress Cataloging in Publication Data

Yellowitz, Irwin.
    Industrialization and the American labor movement,
1850–1900.

    (National University Publications)
    Bibliography: p.
    Includes index.
    1. Labor and laboring classes—United States—
History. 2. Machinery in industry—United States—History.
3. Trade-unions—United States—History. I. Title
HD8072.Y39          331'.0973          76-23090
ISBN 0-8046-9150-9

# CONTENTS

# ACKNOWLEDGMENTS

During the years that I have worked on this book, I have had the good fortune to receive help from many sources. First, I deeply appreciate the grant from the Weintraub Foundation, which was indispensable at a crucial point in the project. Without this support, the volume would have been long delayed. I would also like to acknowledge the help of the City College Research Committee, which met some of the costs of preparing the manuscript. Bernice Linder once again typed the manuscript in exemplary fashion even though she faced more than the usual problems. The librarians at Johns Hopkins University and the Wisconsin State Historical Society were particularly helpful over extended periods of time. Finally, my wife, Barbara, deserves much credit for her constant encouragement and support throughout these years.

# INDUSTRIALIZATION
### and the
## AMERICAN LABOR MOVEMENT
## 1850-1900

# INTRODUCTION

The effects of industrialization are continuous, and today Americans face a new level of technological change as a result of automation. However, current developments take place in a highly organized industrial society that has had over a century of experience with technological innovation. In societies that have social structures and attitudes which are largely preindustrial, technological changes alter the very fabric of life and labor. The United States faced such a period during the last half of the nineteenth century, as this nation developed from a basically agricultural society, with minimal industrial activity and mechanization, into the world's leading manufacturing country. In the process many institutions were changed significantly, but certainly the impact of industrialization was critical for American workers. They had to confront the direct issue of displacement by machines, but perhaps equally important were the indirect effects of rapid industrial change. Thus unemployment assumed new importance; the volume of production suddenly seemed too great—a concept hard to accept when one remembered man's agelong experience with scarcity or considered the reality of continued poverty in the midst of industrial growth; the new factories and mines attracted greater and greater numbers of immigrants, whose impact upon those already employed seemed negative; the hours of labor became vitally important, not only as a means of easing the burden of toil, but as an antidote to the effects of industrialization; and the very structure of society seemed altered by developments that downgraded skill, broke the connection between skill and entrepreneurship, and generally converted the American worker from a person of worth, whose social mobility was assumed, to a dependent wage earner, fixed in a relatively closed working class.

How did organized workers, with the skills and expectations of a preindustrial society, react to economic changes that virtually reconstructed the nation and the role of labor within it? The characteristics of the emerging industrial order were not yet clear, and the reactions of such workers reflected their most basic feelings about job security, personal worth, and economic opportunity. The historian can discover the factors that led these workers to choose one avenue of action over another and can assess why particular policies succeeded or failed.

A study of the alternatives chosen, and of those rejected, leads to a fascinating set of issues. The extent of opposition is an important question. One can understand organized labor's interest in the eight-hour day only in relation to the seeming ineffectiveness of other attempts to meet the challenges of industrialization. The skilled worker's hostility to the immigrant and to female and child labor clearly flow from the competition for jobs in a system that created vast numbers of unskilled and semiskilled places at the expense of the trained craftsman. The relationship between apprenticeship regulations and technological change must also be examined.

Another set of questions concerns the allies that organized labor might find in its efforts to modify the effects of industrialization. Why could trade unions not cooperate more effectively with small companies, which also faced severe competition from machinery? Were reformers and labor leaders able to agree on the present effects and future possibilities of an industrial system? A failure to develop allies would leave labor leaders with fewer viable alternatives.

By 1850, the craftsman already faced division of his work, and the artisan was often a worker skilled in making part of a product rather than the entire item. In the 1860's, labor leaders spoke with envy or nostalgia of the independent master craftsman—working in his own shop with a few journeymen and apprentices and essentially producing a product with a minimum of machinery. Even without extensive mechanization, labor leaders of the 1860's already recognized the impact of division of labor, the increasing size of the workshop, and the tendency to concentrate production in factories. They understood the results of these developments upon the material interests and status of the artisan, who was more likely to become a permanent wage earner than a small entrepreneur.

The American labor movement responded by seeking to apply the principles and tactics it had developed in a preindustrial situation to the changes produced by industrialization. The most basic of these practices was the effort by workers and their unions to control the supply of labor in a trade as fully as possible. From this control, or lack of it, flowed the trade union's ability to modify conditions in the interest of the workers. Almost every major concern of workers, including wages, hours, and, of course, the strength of the union itself, depended ultimately upon such control.

Trade union leaders believed that employers and workers were subject to the classic supply-and-demand situation that gave advantage to the supplier of any product or service only when the demand exceeded the supply. Thus their objective was to encourage those developments in the American economy that limited the supply of workmen, and to create, wherever possible, specific institutions in a trade that would further regulate the number who offered their labor.

Industrialization threatened to destroy the efforts of trade unions to regulate supply, instead creating conditions which would greatly increase the number seeking employment as against the demand for workers. Much of organized labor's response to the machine, the factory, and the other elements of an industrial order was designed to prevent this result. Trade unions realized that the greatest danger from the machine was the destruction of skill, which had been the worker's major weapon for controlling the supply of labor in the preindustrial economy. Competition among employers for competent craftsmen naturally enhanced their bargaining position, but equally important was the relative scarcity of such workmen. The supply was limited by the need to have those already skilled at the trade instruct newcomers, which provided the workers with considerable influence over who should be taught and in what number. Although employers wanted as many workers available as possible—and therefore demanded a large number of apprentices—they were limited in their ability to increase the supply of labor by the small size of their businesses and the need to depend upon the journeymen for the training of young workers.

Employers were eager to develop methods of production that would bypass these limits on the supply of skilled labor. Through division of labor, many crafts had already been significantly altered by 1860. Mechanization only intensified the substitution of semiskilled and unskilled workers for craftsmen. The process was a continuous one as the new techniques, tools, or machines progressively reduced skill and thus increased the supply of workers available. In addition, expansion of markets gave the employer reason to need the expanded production that the innovations yielded; new developments in the financing of corporations provided the larger sums necessary to build factories; and an increased supply of workers from abroad and from the rural areas of the United States provided the labor to operate the machines.

The total effect of these changes was to shift the balance toward the employer in the bargaining process as the skilled worker found that the older methods of limiting the supply of labor became increasingly ineffective. In some cases, the craft faced destruction as the machine eliminated the need for skill; in other cases, some skill was still needed for the most efficient operation of the machines, and thus craftsmen might still be employed in preference to unskilled workers. Even in this situation, the degree of training required was usually less, and thus the ability of the trade union to control the supply of labor was reduced.

A second basic objective of workers and their organizations, which continued from the preindustrial into the industrial period, was protection against unemployment. This was related to the supply of labor but was also independent of it. A general depression in the economy or seasonality in a trade could throw men out of work, thereby increasing the supply of workers available and still further modifying the bargaining situation in favor of the employer. Yet limits on the labor supply did not necessarily protect workers from unemployment, which might result from factors entirely within the control of the employer. Workers might have some control over the teaching of a skill and the number of persons who might be available for work, but the employer had absolute authority on whether they should work at all. Trade unions frequently charged that this power of discharge was used by employers to create an underemployed group of workers who would exert a negative influence on the conditions of those still at work. Fear of an increase in the number of apprentices was motivated, in good measure, by the knowledge that the newcomers would be unlikely to have full-time employment.

Labor leaders argued that unemployment became endemic in the last half of the nineteenth century because the number of those seeking work increased faster than new jobs became available. In effect this meant that workers had to share jobs. Moreover, they claimed that the periodic business crises were largely the result of overproduction in mechanized industries. Until the overstock was gradually absorbed into the economy, unemployment was substantially increased. In many trades, seasonality was ascribed to the same mechanism in miniature. The surfeit of workers and goods prevented the full employment of those who sought work and the sale of all the goods produced.

The response of organized labor was to modify older devices in order to make them more effective as limits upon the supply of labor and goods. After 1873, the call for the eight-hour day emphasized the effect of shorter hours in curbing overproduction. The hostility to certain groups of immigrants in the 1860's and early 1870's was broadened during the last two decades of the century into a call for significant limitations on entry into the country. Trade unions also attempted to limit production itself in an effort to control seasonality and the underemployment of the work force. In the past, trade unions had sought such limits to prevent overwork, and because of the persistence of some variety of the idea that the total amount of employment was a fixed quantity that had to be stretched over time and divided among the existing workers. However, these reasons clearly decline in importance as the century advances, and the emphasis shifts to preventing overproduction rather than preserving a limited quantity of work. Industrialization seemed to create the anomaly of too little work because of too many goods.

Labor leaders were well aware of the advantages of industrial progress for the consumer, but again long-standing principles from the preindustrial period asserted the primacy of the worker as producer rather than consumer. Workers believed that a higher standard of living resulted from higher wages even if this increased the cost of production. Low prices and low wages were of no value to the worker. In the 1870's, this concept became the basis for an underconsumption argument that paralleled the emphasis on overproduction. Relying heavily on the theory of Ira Steward, organized labor and its supporters argued for the advantages of a high wage system. Without adequate wages and the resulting ability to consume, industrialization meant glut and depression. Since the general community suffered in such a situation, the public had a concrete reason for supporting the labor movement's demands for decent wages. Labor leaders then made the connections to specific conditions in their trades that seemed to generate low wages.

Tactics varied from industry to industry, but one can delineate two fundamental developments. First there is no succession in tactics over time. Unions in the 1890's had much the same approach as those in the 1860's. Disputes within a union over policy were as likely to develop at the end of the century as in the 1860's, and such conflicts were usually the result of the different interests of workers in the same union, as technological change modified conditions for some more quickly than others. Second, skilled workers and their unions did not usually try to obstruct technological change. Clearly some skilled workers, particularly in the early years of a significant innovation that affected them directly, believed that opposition was a viable policy. Yet at the same time, other skilled workers demanded some type of accommodation, and this group usually prevailed within the trade union. However, one must be careful not to equate such accommodation with acceptance of technological change in the form and at the rate that an employer would most desire. Implicit in organized labor's acceptance of innovation were steps to protect the interests of the existing work force to the extent possible. Uncontrolled technological change was not the sole model for innovation, but rather the one most favored by the innovators. Despite the widespread emphasis on progress and the universality of benefits, trade unions claimed that unrestrained innovation actually served the interests of the few at the expense of the existing work force and the entire society. Everyone suffered from the overproduction, recurrent economic crises, and unemployment that labor spokesmen attributed to uncontrolled technological change. Instead organized labor suggested an alternative that recognized the inevitability of change, but also sought to reconcile innovation with the interests and needs of the established society.

Basically, organized labor refused to define progress independently of the social and economic context. The interests of specific capitalists dictated whether technological changes would be introduced and determined the subsequent rate of development. Workers had the same right to insist on the protection of their interests. Accordingly, existing relationships should be changed only with a clear understanding of the effects that were to follow, and with attention to the interests of those in the society. Instead of an unrestrained economy, responding to untrammeled self-interest, committed to the victory of the stronger, and guided by a series of supposedly natural and unchangeable principles, economic and social arrangements should be modified in order to serve the recognized interests of clearly defined groups such as the trade unions. A lack of such restraint would inevitably lead to increased poverty, sustained unemployment, social disruption, and personal degradation.

By the close of the nineteenth century, organized farmers, through Populism, and a growing body of reformers had come to the same general conclusion as the labor movement. Thus an effort to conserve existing interests stimulated the growth of American reform. The state was inevitably a means toward this end. However, it became less important to organized labor than to other groups that demanded restrained industrialization. Though the labor movement was prepared to use the state as a major mechanism of regulation in areas such as female and child labor and immigration, and though political action was a constant feature of the American labor movement, the trade unions became the primary instrument for meeting the impact of industrial change upon workers. In part, this emphasis on the trade union flowed from the failure of political action in the period from the 1830's to the 1870's. The major political efforts of labor leaders in the 1860's had produced little, and the trade unions of the 1880's were less willing to place emphasis on the state as the means of meeting the problems faced by workers. Action by trade unions did maximize the labor movement's control over what action should be taken, but at the same time it limited the range of available options and reduced the possibility of an alliance with other groups which relied upon state action to check uncontrolled industrialization.

Despite the differences over means, it would be a mistake to underestimate the impact of the labor movement upon the development of the fundamental policy of restrained industrial capitalism that ultimately became the course of American reform. Though organized labor lacked important theoreticians, and though it usually expressed major positions through specific issues, one should not miss the fact that there was a clear ideology. Contemporaries did not fail to see the labor movement's basic attack upon the principles and procedures of unrestrained industrialization. To understand how the American labor movement proposed to meet the challenge of industrial change, and to discover how other groups reacted to organized labor's positions and actions, is the purpose of this volume.

## CHAPTER ONE

# THE POSITIVE VIEW OF MECHANIZATION:
## The 1860's

In the period to 1873, the major elements of organized labor's response to industrialization are clearly present even though machinery still had a relatively limited effect upon skilled workers. In trades that were already significantly affected by mechanical innovation, principally shoes, cigars, and cooperage, proponents of all-out opposition contended with those who called for accommodation on terms that would cushion the impact on the skilled worker as fully as possible. However, this was also the period in which those who viewed the machine as a boon to the worker had their widest hearing within the labor movement. Many leading spokesmen argued for the positive effects of machinery upon workmen. Among these were Jonathan Fincher, a leader of the Machinists' and Blacksmiths' International Union and editor of the major labor newspaper published during the first half of the 1860's; Andrew Cameron, editor of the leading labor newspaper of the late 1860's and early 1870's, and a major figure in the National Labor Union; William Sylvis, the prominent President of the Iron Molders' International Union and a president of the National Labor Union; Thomas Phillips, a well-known cooperationist, and a leader in the shoemakers' union—the Knights of St. Crispin; and Ira Steward, the influential theorist and leader of the movement for eight hours. Together they provided the most wide-ranging and influential defense of machinery that was to emerge from within the labor movement in the nineteenth century.

In part the proponents of machinery merely accepted a basic fascination with progress. The editor of *Harper's Monthly* spoke for much of American opinion when he lauded the progress in the "mechanic arts," suggesting that such advances were bound to reach the poor even if innovation at first seemed to increase pauperism. Ultimately he connected the amazing march of invention with some providential plan, in which the United States was apparently chosen to play a major role. Invention would unite with commerce and religion to determine mankind's future. The editor concluded that machinery would lift the burden of toil from the backs of the slaves and all mankind: "We who have no slaves have machines in place of them."[1] President Robert C. Smith of the National Typographical Union discussed advances in the technology of communication at the union's convention in 1858. He welcomed inventions, such as the power-driven printing press, which were bound to increase intelligence among the people.[2] Andrew Cameron's *Workingman's Advocate* was one of the major labor newspapers of the period, and it contained regular columns on scientific and technical advances.[3] Cameron complained not of machinery, but that the workmen who invented it did not receive the esteem or income given doctors, lawyers, or merchants, although their innovations were revolutionizing the world and serving the public interest. Cameron apparently saw no reason to fear that the skilled artisan might successfully innovate his craft out of existence. Instead, the threat to the skilled worker came from low wages and a lack of recognition, which dissuaded the talented artisan from remaining in a trade while encouraging the proliferation of "botch" workers, who could not even do proper work let alone invent something to benefit society.[4]

William Sylvis and Jonathan Fincher were among the most influential labor spokesmen in the 1860's, and both gave strong support to innovation. Sylvis argued before the convention of the Iron Molders' International Union that the condition of labor would not be improved by restricting production, and anything that blocked the salutary effects of machinery on production reduced the total wealth of the world.[5]

Jonathan Fincher admitted that in the past, invention had usually created fear and anxiety among workers, but he also made it clear that he "never entertained the fears, experienced by many, of the probable or possible evils that might be entailed upon us from the adoption of labor-saving machinery." In fact, Fincher also saw the beneficence of a divine being in the unparalleled comfort that invention had afforded the producing classes. He joined President Smith of the typographical union in his praise of the positive benefits of the power-driven printing press—an example surely to be expected from the leading labor editor of the period. But what of the displacing effect of the machine? Fincher recognized that an invention might drive ten or twenty men from employment, but he insisted that it "would open the door, in the same establishment, to fifty or one hundred additional hands to be employed in some other department."[6] Thus he joined almost every proponent of mechanization in stressing that machinery displaced men only in the short run since it created many more jobs than it destroyed. However, Fincher was silent on whether the jobs created were filled by the men displaced, and if so, whether the conditions of employment in the new job compared favorably with those of the one lost. Labor leaders in trades affected by the machine were aware that the answers to these specific questions were more important to the fate of the worker involved than any general thesis that was obviously true for the economy as a whole but not necessarily for the existing skilled labor force.

Fincher joined Sylvis in emphasizing that wages were dependent on the total size of the product available as well as the way such a product was divided.[7] Trade unions were designed to insure as equitable a division as possible, but the pie must be there in the first place. Machinery also allowed production of goods that were once luxuries in sufficient amounts to reduce prices and bring them within the reach of the workmen. Finally, the machine lessened physical toil, although overwork was still a real danger if the machine, and not the worker, set the pace of labor. The intensity of labor required in working with machines was one more reason why the hours of labor had to be reduced.[8]

At no time in the subsequent course of industrialization in nineteenth-century America did so positive a view of the machine emerge from major figures within the labor movement. A number of factors explain the situation, but the most important was the limited impact of the machine in the 1860's. Although the manufacture of shoes, barrels, and cigars underwent significant change in the 1860's and early 1870's, machinery displaced few skilled workers outside these trades. For example, President William Saffin of the iron molders wrote in 1875 about the lack of improvements in the foundries.[9] Thus Sylvis did not have to face the choice posed for the shoemakers, coopers, or cigar makers as they contemplated innovations that increased production and reduced price, but also eliminated skill and converted artisans into members of a production force rather than producers themselves.

The introduction of machinery might supplant the unskilled as well as the skilled. For example, the introduction of equipment in New York harbor in 1861 replaced hand labor for loading grain; this displaced thousands of laborers within the next few years. Despite the demand for all types of workers during the Civil War, two thousand laborers struck against the use of the machinery in 1862. The strike was lost, but during the draft riots in New York City in 1863, some of this equipment was destroyed by the shovelers.[10]

Much of the machinery praised by labor spokesmen simply substituted mechanical for hand energy, which assisted the craftsman without eliminating his skill. In this situation, skilled workmen could share in the benefits of machinery in the same fashion as other elements of society. The machine lightened labor, produced more wealth, and lowered prices, which brought goods within the reach of the laboring family. In addition, Fincher was a machinist, and the trade prospered from the introduction of machinery produced by these craftsmen. Fincher noted that the need for cotton, railroad, and marine machinery in the 1850's had developed the trade.[11] In the crafts in which machinery was displacing skill itself, labor leaders failed to share the enthusiasm of Fincher and Sylvis.[12]

It is also important to understand that proponents of machinery did not regard innovation in general as a blessing. Thus the rapid division of work was a more significant threat to skilled labor at this time than the machine.

Sylvis and Fincher were hostile to the breakdown of apprenticeship that accompanied the dividing of tasks once done completely by a skilled workman alone or with a small team of helpers.[13] The division of work made it possible for apprentices, with some training, to do many more of the tasks required in production. Their presence increased the total supply of workers in a trade without reference to the actual demand for labor. Since most skilled workers believed that the relationship between the supply and demand for labor largely determined wages, the apprenticeship issue was so serious that it stimulated the formation of several trade unions in the 1850's, including the printers, machinists and blacksmiths, and iron molders. Every trade union in the period sought to limit the number of apprentices as a means of increasing its control over the supply of workers. Fincher and Sylvis thus responded to the explicit threat that existed in the form of apprenticeship but ignored the implicit threat that accompanied the machine. They realized the negative effects of too many apprentices because that tangibly increased the supply of labor, but the machine at the time did not put pressure upon the worker as a producer and conferred many benefits to the worker as a consumer.

The Civil War intensified the threat to traditional limits on the number of workers in a trade. The wartime need for greater production led employers to demand that restrictions on the number of apprentices be waived. Jonathan Fincher noted that some journeymen opposed any increase in the number of workers for fear that the supply of workmen would not diminish once the war had ended even though the demand for goods would certainly fall. The result would be an oversupply of labor, which would only be worsened by the return of the troops.[14] Division of labor threatened to destroy skill, but even an increase in the number of craftsmen working in a traditional manner could be a danger if all were not able to secure steady employment. Workmen remembered the panic of 1857, with its burden of unemployment. Even in prosperous years, there were dips in demand, often seasonal, that cut the need for workers and led to a competition among journeymen that served the interests of the employer. Trade unions had been formed to protect the worker against the negative effects of the capitalist economy, and an oversupply of labor was perhaps the most serious threat since trade union leaders so heartily accepted supply and demand as the determinant of wages and most other conditions of employment.

Capitalism did not allow for a harmony of interest between workmen and employers. As William Sylvis put it: "Capitalists employ labor for the amount of profit realized, and workingmen labor for the amount of wages received. This is the only relation existing between them; they are two distinct elements, or rather two distinct classes, with interests as widely separated as the poles. We find capitalists ever watchful of their interests— ever ready to make everything bend to their desire. Then why should not laborers be equally watchful of their interests—equally ready to take advantage of every circumstance to secure good wages and social elevation?"[15] The war threatened to aid employers by providing an excuse for increasing the supply of labor without any guarantee that the demand necessary to employ the larger work force steadily would continue beyond the end of hostilities. Thus Jonathan Fincher made it clear that he favored an increase in the supply of workers for war production only if hours were cut, without a loss in pay, to prevent unemployment once demand slackened.[16]

Fincher and Sylvis also recognized the threat that immigration constituted for the skilled worker, since once again it threatened to increase the supply of labor, thus weakening the bargaining position of the craftsman and setting worker against worker in a destructive competition.[17]

A positive view of the machine did not mean that the dangers attending its use were ignored. Thus Fincher warned that machinery could lead to overproduction. In Fincher's view, overproduction resulted from an increased ability to produce combined with the employers' inability to regulate the supply of goods to the needs of the market because of the necessities of competition. "That competition is so strong, one trying to outdo the other, that employers are powerless to apply any effective checks." The result was industrial panic, as in 1857, and yet employers resisted the efforts of unions to restrict apprentices, which would have helped keep supply in balance with demand by slowing the division of work that allowed for an increased production.[18] The employer feared a glut in the market, but also wanted the competitive advantage that came from increased production and lower cost per unit. The result was "closed mills, cold and cheerless furnaces, silent factories, quiet shops, mute anvils, idle workmen, ill-fed and worst clothed children, wan and anxious looking mothers. . . ." These unfortunate conditions were "caused by *overproduction,* which, in time, produced a commercial and financial crisis and crash, in which the producer suffers far greater distress than the middle men or business men."[19]

Fincher understood that machinery allowed three men to produce more than five had twenty-five years before, which naturally meant a vastly increased supply of goods that might well exceed demand. These unfortunate possibilities could be offset by hours reduction and apprenticeship regulation, which would reduce the tendency to overproduction. He recognized that these might be only palliatives, since "In time, by the introduction of improved machinery, it [overproduction] might be as bad as now; if so, let the men of that day, with the leisure they would enjoy, devise their own remedy, sufficient it is for us to attend to our present duties, and not anticipate the evils of another generation."[20]

Labor leaders also recognized that past attempts to block mechanization had generally failed. However, this generalized understanding was no guarantee that a specific threat against one's own trade would be viewed in light of history.

Finally, organized labor's basic acceptance of property rights and private enterprise made it difficult to challenge the introduction of the machine in any direct fashion, since this was clearly within the employer's right to use his capital as he saw fit. An accommodation to the effects of machinery often avoided any direct confrontation over basic property rights, since the measures proposed concerned apprenticeship, hours, immigration, female and child labor, or work rules—all of which could be regarded as proper subjects for governmental action or agreements between employers and unions. Much of the criticism of the wage system from within the labor movement was based not upon hostility to private property or private enterprise, but upon a desire to stop the changes that seemed increasingly to limit the skilled worker's participation in the system. Whether labor leaders accepted some millennial objective (such as producers' cooperation), concentrated on the protection of workers within an industrial society that would remain intact for the foreseeable future, or still longed for the supposed mobility and status of a preindustrial period, there was little direct challenge to private property and capitalism as such. Thus a theoretical basis for an attack on machinery was lacking.

Still another source for the acceptance of the machine was the eight hour movement, headed by Ira Steward. His contentious nature and rigid insistence on the sufficiency of eight hours—as the means for immediate reform as well as ultimate social reconstruction—increasingly alienated him from other elements in the labor movement.[21] Yet shorter hours was the major political objective for the labor movement in the 1860's, and Steward's influence in this area was preeminent. Thus his views received wide attention.

In contrast to most economic thinkers of the time, Steward did not stress increasing productivity per man-hour and a general increase in total production as the basis for prosperity and progress. Rather, he stressed consumption as the fundamental element in all economic analysis.

Unless the working classes are *paid* sufficient wages, they will not be able to *buy* certain articles which manufacturers and merchants are so eager to sell. Capitalists remember us as *Producers,* to be paid as little as possible; but not as *Consumers, to be paid enough* to enable us to *buy* their commodities. They remember that Foreigners *work* cheaper than Natives; but forget that Natives *buy* more of their goods than Foreigners.

When will they learn that their immediate and special interest in *Cheap Labor* has blinded their eyes to their final and general interest in Labor sufficiently well paid to buy all that they *desire* to manufacture?[22]

As a result of the insufficient consumption that stemmed from insufficient wages, the worker was not only plagued by an inadequate income, when at work, but by recurring periods of idleness. Such unemployment lowered wages still further by increasing competition among workmen for employment; but ultimately the gains to the employer in this regard were more than offset by the financial convulsions that resulted when "Wealth is more rapidly produced than consumed." Contrary to popular belief, "Instead of bankruptcy being the cause of enforced idleness, enforced idleness is the cause of bankruptcy." Unemployment could be relieved by increasing wages, which would stimulate demand and thus create employment for more workers. This would not only increase wages by eliminating the oversupply of available labor, but prevent panics and depression by insuring a continuing demand for the products of industry. Steward stated again and again that "dear men are the cheapest, and cheap men the dearest."[23] Thus the employer and worker basically had an identity of interest: low wages ultimately hurt both by cutting consumption and producing depression, whereas high wages were the basis for a steady demand that would mean enhanced profits for the employer.

Steward gave much attention to the factors that determined wages, and he quickly discarded the position of the trade unions. Thus control of the supply of labor might be effective in the short run in winning higher wages for skilled workers, but it would do little for the work force as a whole. Secondly, even skill itself was no guarantee of high wages: witness the low pay of craftsmen in nations such as Japan and Belgium.[24] Steward did not deny that traditional trade union policies, based on the maintenance of skill and the control of the supply of labor, could have a short-run effect, but ultimately they were overwhelmed by other, more potent factors, of which the most crucial was the level of wants or desires of the mass of the people.

To Steward, wages were basically set at the level demanded by the standard of living of the nation at a particular point in its history. The supply of labor, availability of capital, and the activities of trade unions were of secondary importance. Employers paid the wages demanded by the level of wants of their workers, which, in turn, were set by the habits, customs, and expectations of the employees. Wages were low in the Orient because the standard of living was low. Thus history ever played out its influence upon the present generation. Higher expectations among American workers set wage levels above those of Europe, but this standard of living was still too low to absorb the production generated by American industry. In order to improve the standard of living, and thus wages, Steward relied on shorter hours, specifically the eight-hour day. Shorter hours would allow men the time and energy to cultivate their tastes, acquire new interests, and enjoy more of what they already desired, with the inevitable result that as the level of wants and desires increased, wages would rise accordingly.[25] A shorter work day was important to Steward precisely because it generated a new level of wants, which, in turn, was the key to wages and the growth of industry to meet the new demands generated by a higher standard of living.

Steward's theory also led him to deny the basic American virtue of thrift. He agreed with labor spokesmen who felt that the decline of skill and small business in the face of the division of work, machinery, factories, and corporations made it unlikely that craftsmen would advance to become small entrepreneurs. Once saving for capital accumulation became a useless objective, thrift lost any value for Steward and actually became a negative factor in determining the level of wages.

Suppose for example that the average pay is a dollar a day; and that every laborer cuts his outgoes down sufficiently to save one-quarter of his income. Wages would soon slide back from the dollar formerly expended, to the seventy-five cents necessary to pay for the reduced expenditures. All that can hold wages up to a dollar is the fact that an average laborer expends a dollar.[26]

The highest consumption possible was always the goal.

Of course, a higher standard of living could only become a reality if production could be expanded to meet the increased demand. The machine was the vehicle for accomplishing this increase in production, and it was thus basic to Steward's entire system. In the process, he turned his back on skilled hand labor and thus on the membership of the trade union movement.

In the rapid production of wealth, very little can be done by hand labor, or with tools that require about all of the strength and skill of the human being to use them. The quality of work done with such tools is sometimes excellent, but a great deal of time and labor is required to produce a very limited quantity.[27]

The machine produced the wealth necessary to fulfill the increased demands that would result from an eight-hour day. Hand labor meant restricted production, and by necessity poverty; machinery meant almost unlimited production, which was the necessary condition for an increasing standard of living for the masses of the people.

There was no glorification of the artisan or the small shop. Instead the prominent cooperationist and labor leader Thomas Phillips pointed out that hand labor had meant long hours in order to satisfy a lower standard of living, and that the goods produced were much more expensive than those made by machinery and division of labor. Machinery allowed employers to pay good wages and cut prices, something that was impossible when hand labor dominated industry.[28]

Ultimately Steward had to face the same problem that confronted all advocates of mechanization: how does one meet the tendency of machinery to displace men in the present or future? Steward offered nothing more on this basic question than many others. He admitted that machinery would destroy old industries based upon hand labor, and in his thinking this was not only inevitable but laudable. "Machinery is successful or saves labor, just in proportion as it discharges those who were formerly indispensable. The more it discharges, or the more employments it destroys, the grander are its achievements."[29] However, there will not be permanent displacement of workers so long as hours are reduced and wants therefore increased, since the higher standard of living will call into being new industries and many new jobs in older trades in order to meet the increased demand.[30]

Yet the general increase in employment, which would occur over time, did not necessarily meet the immediate problems of those displaced from hand labor. Would they secure the new jobs, and if so, at what wages? Although wages for the entire work force would rise once hours were reduced, there would necessarily be a time lag, and there would still be gradations of wages within the labor force. Would skilled workers have to accept the lower wages of a machine tender in the period of displacement? Although Steward accepted a decline of wages for those initially displaced, he emphasized the general situation, which would be an upward movement of wages. In this regard, he joined both traditional economists and other reformers, who praised mechanization for its general and long-term effects but offered little on the short-run issue of how the present work force was to contend with the admitted displacement that would accompany the mechanization of handwork. Yet the short-run issue was the key for skilled workers and for union leaders in those trades affected by mechanization.

One possible solution would have been action by the government through a slowing of innovation or by compensating the workers displaced. This required principles of social interdependence and community responsibility that were lacking in the competitive economy of the period. Not until the early twentieth century was unemployment faced as a basic issue in Europe, and action in the United States awaited the depression of the 1930's. Thus it was up to the trade union to act as the sole buffer against the unregulated effects of industrialization.

Steward was aware of labor's hostility to machinery in the past, but he believed that his writings would convince workers and trade union leaders of the essential importance of machinery to a higher standard of living. However, the worker was not the only enemy of the machine. Although capitalists seemed to be the proponents of mechanization, Steward charged that they could, in fact, be the most serious enemy.

Capitalists generally conceived of machinery as a means of reducing wages, which Steward believed was sure to impede mechanization in the long run.[31] Machinery was used when labor was dear, and thus he had no quarrel with the employer's desire to reduce the cost of production. Yet employers confused wages and cost of production, and Steward insisted that costs could be lowered, through the vehicle of mass production and mass markets, while wages were increased. If wages were cut, demand necessarily fell, and the introduction of machinery was slowed since the required demand was not present. Moreover, cheap labor blocked innovation: the employer will only introduce machines if he sees the possibility of reducing costs. "In China, sedan chairs are used to carry passengers instead of horses and carriages. Not because transportation upon men's shoulders is easier, or more rapid and pleasant than by horse power; but because there, a man undersells a horse."[32]

Steward noted that cheap labor abroad undercut wage standards in America.[33] This would inhibit mechanization by substituting cheap human beings for the cheapening power of the machine. Thus if mechanization were not to be slowed, Steward believed it necessary to equalize the wages of workers worldwide.[34] However, during the last half of the nineteenth century, the United States experienced both the replacement of expensive skilled labor with cheaper unskilled workers—often from abroad—and massive mechanization. Steward did not believe this to be possible, because he thought exclusively in terms of the increase of demand through higher wages while ignoring the effects of an absolute increase in the number of consumers, many of whom were foreign born.

As we have observed, Steward argued that the relatively high wages for

American skilled labor had to be sacrificed temporarily because hand methods would not meet the needs of a higher standard of living for the mass of the people. Yet it was quite conceivable that even though production increased as a result of mechanization, the habits and desires of the populace would remain stable because their hours were not reduced. Steward recognized that in such a case the condition of all workers would worsen. Unemployment would increase, and ultimately goods without buyers would create depression and general economic stagnation. Thus the eight-hour day was of absolute importance to Steward as the means of increasing the level of wants and avoiding the glut which would otherwise accompany the mechanization of industry.

Steward's fascination with the machine ultimately led to technology becoming the agent of the social reconstruction that was to be the end product of the movement for shorter hours. High wages would make every laborer on earth "sufficiently a capitalist," but also would force continuous mechanization as capitalists sought to cut costs. The result would be that the cost of such improvements would bankrupt even the wealthiest capitalist, creating a single class formed from the former laborers and capitalists. "And when every human being includes within himself, the interests and duties of these two classes, the conflict between them will cease; for under such circumstances a contest between a laborer and a capitalist would mean a man contending with himself."[35]

Combined with the support of the machine by labor leaders as prominent as William Sylvis and Jonathan Fincher, the position of the Stewardites represents the high point of the acceptance of the machine by the American labor movement. Even in the 1860's, some opposition appeared among workers and trade union leaders who directly confronted new machinery. In succeeding decades, the labor movement regarded the machine as an agent of destructive change rather than a major element in the progress of the nation. Although few trade unions in the 1860's, or later decades, adopted a policy of opposition, labor leaders who supported accommodation to technological change lacked the positive attitude toward machinery so prominent in the 1860's.

## CHAPTER TWO

# INDUSTRIALIZATION AND UNEMPLOYMENT

Although technological innovation in the early 1870's was modest by the standards of the last two decades of the century, it was already significant enough to heighten fears of the machine within the labor movement.[1] Labor leaders increasingly stressed the negative effects of industrial change upon employment and wage levels. In 1872, workers in more than thirty trades in New York City took part in a strike for the eight-hour day. At one of the mass meetings held during the height of the strike, a resolution was adopted that firmly linked the eight-hour day with the increasing use of machinery. The resolution argued that machinery already performed much of the work formerly done by simpler methods, thus diminishing the need for labor. As a result, the present needs of society could be met without full employment. Unless hours were reduced, mechanization would create either seasonal unemployment or a permanent oversupply of workers. This situation meant that only employers had gained from the introduction of machinery, since they saved the price of the displaced workers. The resolution demanded that workers also be permitted to share in the benefits of mechanization through an immediate reduction in hours to eight per day.[2] Labor leaders believed that such a reduction would allow more workers to be employed to manufacture the existing volume of goods. A shorter work day would soak up many of the unemployed and remove the deadening pressure of the idle worker upon the ability of those employed to secure the best wages and terms of employment.

The depression of the 1870's intensified the fear of machinery. The optimism about the effects of mechanization, so prominent among labor spokesmen in the 1860's, vanished. There was almost universal agreement in the labor movement that "The vast increase in the producing power of machinery has resulted in a decrease in the number of persons required to keep up the supply of any given manufacture in which machinery is employed, so that a large number of persons are at any given time unemployed."[3] This position denied the widely held contention that the unemployment of the 1870's was a result of the panic of 1873 and that it would disappear once business improved. If machinery were the basic reason for the existing idleness, unemployment would not end with the revival of business but would continue as a permanent feature of American life.[4]

Labor journals dropped their columns on mechanical improvements, and attacks upon the machine as the basic source of the workers' distress appeared more frequently.[5] Even supporters of technological innovation had to acknowledge that machinery "has been a seeming evil to the thousands it has deprived of employment and bread. . . ."[6] Though still a supporter of Ira Steward's eight hour theory, George E. McNeill dropped the praise of machinery, so central to Steward's thesis, and instead stressed that industrial change was discharging men more quickly than new employment could be opened for them. Only eight hours could offset this destructive development.[7] The Centennial Exposition of 1876 in Philadelphia focused attention on the new "industrial age," but also revealed even more clearly a lack of any governmental policy to offset the poverty and depression that seemed to accompany the increasing productiveness of labor.[8]

The effects of division of labor and the breakdown of apprenticeship had stimulated the formation of trade unions in the 1850's and 1860's, and the Council of Trades and Labor Unions of Chicago believed that the rapid displacement of work by machinery had to revitalize the labor movement of the 1870's, so shattered by the long depression that followed the panic of 1873. Unless this were done, a permanent condition of mass unemployment would result. Labor organizations were urged to include eight hours as part of their constitutions.[9]

Although mechanization still affected only a minority of the labor force in the 1870's, labor leaders believed that the machine had become an unqualified threat to skilled workers. As the use of machinery expanded in the 1880's and 1890's, the labor movement increased its effort to appraise the effects more fully and develop appropriate defenses. Nonlabor observers also stressed the connection between mechanization, unemployment, and social discontent, and they sought serious information about the extent and impact of technological change.

Labor leaders insisted with increasing frequency during the 1880's and 1890's that unemployment was primarily a result of mechanization, and unless hours were reduced, a major social problem would result. The effects of machinery were discussed at length during the hearings on industrial conditions held in 1883 by the United States Senate Committee on Education and Labor.[10] There was general agreement among the labor leaders who testified that mechanization already had altered the system of production built on skill and small shops. Displacement of workers and an overcrowding of the labor market were real dangers which required the immediate establishment of a shorter work day.

The American Federation of Labor consistently pursued the same objective. The eight hour movement of the 1880's and early 1890's was one of the major activities of the federation, and it was clearly based upon the link between unemployment and machinery. The executive council put the issue squarely in 1888 in a letter to the International Trade Union Congress. Although denying any intention to limit production, so long as any person was lacking food or shelter, the council made it clear that its primary concern was the producer, who faced the destructive results of technological innovation. Thus "in view of the fact that the application of steam machinery and the minute sub-division of the processes of industry are continually throwing large numbers of our fellow [workmen] out of employment, and that the permanently unemployed class is rapidly increasing—so that in this country, where the natural opportunities are so vast, more than five percentum of the population are doomed to a fate worse than death—we should make a strenuous effort to reduce the hours of labor to such a point as would afford to all the opportunity to labor; that is to say, to the means of life." Shorter hours could best be secured through the action of trade unions, with the role of the state limited to regulating child labor.[11]

Reports by individual unions to the New York Bureau of Labor Statistics claimed machinery reduced the time needed to produce an item and saved the labor of a specified number of workers.[12] The trade unions said little about the actual number displaced, nor did they indicate how many workers failed to find other jobs in the trade or whether these new positions were skilled or unskilled. Also, there was no comment on whether some of the workers displaced had not found work in other trades. Clearly, most labor unions lacked this information, and thus these statements about unemployment were general in character. They were designed to establish the causal relationship between unemployment and the machine rather than its extent. A variety of governmental agencies sought to fill this quantitative gap.

Massachusetts undertook a state census in 1885, and the enumerators were asked to inquire about unemployment as part of their home-by-home canvass. In 1887, the Massachusetts Bureau of Labor Statistics used the raw census schedules as the basis for a study of the extent of unemployment in the state.[13] For the year preceding May 1, 1885, 70.41 percent of those

TABLE 2.1

**Percentage of Unemployed Persons in Massachusetts, 1884–1885**
(By Number of Months Unemployed)

| Number of Months | Male | Female | Total |
|---|---|---|---|
| One | 7.25 | 10.53 | 8.10 |
| Two | 18.34 | 23.86 | 19.78 |
| Three | 16.87 | 18.65 | 17.33 |
| Four | 21.16 | 15.28 | 19.63 |
| Five | 7.22 | 5.33 | 6.73 |
| Six | 18.66 | 15.05 | 17.72 |
| Seven | 2.58 | 2.44 | 2.54 |
| Eight | 2.99 | 2.91 | 2.97 |
| Nine | 2.09 | 2.51 | 2.20 |
| Ten | 1.59 | 2.08 | 1.72 |
| Eleven | 0.83 | 1.25 | 0.94 |
| Twelve | 0.42 | 0.11 | 0.34 |

Source:  Massachusetts Bureau of Labor Statistics, *Eighteenth Annual Report*, 1887, p. 156.

TABLE 2.2

**Percentage of Work Force Unemployed for Some Portion of the Year Preceding May 1, 1885**
(Selected Cities, Massachusetts)

| City | % | City | % |
|---|---|---|---|
| Boston | 18.40 | Malden | 32.35 |
| Brockton | 55.37 | New Bedford | 22.33 |
| Cambridge | 26.49 | Newburyport | 38.98 |
| Chelsea | 9.34 | Newton | 28.14 |
| Fall River | 56.38 | Northampton | 42.28 |
| Fitchburg | 14.21 | Salem | 23.15 |
| Gloucester | 20.61 | Somerville | 13.71 |
| Haverhill | 41.61 | Springfield | 20.21 |
| Holyoke | 24.75 | Taunton | 42.65 |
| Lawrence | 36.37 | Waltham | 28.21 |
| Lowell | 33.71 | Worcester | 32.66 |
| Lynn | 43.13 | | |

Source:  Massachusetts Bureau of Labor Statistics, *Eighteenth Annual Report*, 1887, p. 277.

queried had worked a full year, while 29.59 percent had been unemployed at some point.[14] Table 2.1 shows the length of time such persons were unemployed.

Unemployment was also spread throughout the state (see Table 2.2). Boston had a lower percentage of unemployed than the state as a whole, and a far lower rate of unemployment than many of the smaller industrial cities. Commissioner Horace Wadlin of the Massachusetts Bureau of Labor Statistics attributed this situation to the variety of occupations present in Boston, which prevented workers from becoming dependent on the fortunes of a single industry.[15]

The census also broke down its figures by percentages unemployed in various occupations. Wide variations appeared, with only 9.91 percent of male printers and compositors unemployed at some time during the year as compared with 39.09 percent for male cotton mill operatives, 47.18 percent for male carpenters, and 67.30 percent for male boot and shoe workers.[16]

These figures gave a comprehensive picture of the extent of unemployment, and they revealed that idleness was a basic feature of America's economic life. The total number and percentage unemployed, the long periods of unemployment, and the obvious importance of seasonality in many occupations raised serious questions about the ability of American industry to employ fully the available work force. It confirmed the arguments of labor leaders that unemployment was a serious menace to the American worker and gave a boost to the call for an immediate eight-hour day.

Some contended that the census figures had to be used with care, since 1884–85 had been a depression year. The Commissioner of Labor Statistics for Massachusetts, Carroll Wright, supported this view by pointing out that the wide distribution of the unemployed was influenced by the "more or less depressed condition, the depression which began early in 1882 not then having passed away."[17] However, as Wright later pointed out, federal census data for 1890 still showed 18.27 percent unemployed in Massachusetts as compared with the 29.59 percent in the census of 1885. In addition, the average period of unemployment was quite similar: 3.74 months in 1890 as compared with 4.11 months in 1884–85.[18] Such variations could certainly be accounted for by the business cycle, if not by the accuracy of the census methods. In 1893, Horace Wadlin insisted that the figures for 1884–85 were fairly representative even though certain industries, such as boots and shoes and cotton goods, had not operated to the fullest. The year was not one of "extensive industrial depression," and it was not unusual at any time to find specific industries that were depressed as compared with other industries or with their own performance in other years.[19]

The Massachusetts census did not attempt to seek the causes of unemployment, but an investigation by the United States Bureau of Labor, under its first chief, Carroll Wright, did focus on the displacement of men by machinery. The results confirmed the fears and warnings of labor leaders. Based upon comments by employers, trade after trade revealed rates of displacement from 25 to 75 percent of the former skilled work force.[20] Although exact percentages would be meaningless in such a survey, since the estimates of employers were not verified independently by the Bureau of Labor, the overall message was clear.

Wright did not indicate whether the men displaced found work elsewhere, either as machine tenders in their original trade, or in other occupations. Such information was difficult to obtain without a follow-up of individual workers that was beyond the resources of the bureaus of labor statistics or the trade unions. It was an important point, since defenders of machinery claimed that any initial displacement would soon be eliminated once mechanization, mass production, and lower prices had increased demand. Only if an entire trade were eliminated would men be driven to seek employment in other occupations; and in such a case, new industries that resulted from technological innovation would more than meet the need. However, reports on individual trades suggested that a significant number of those displaced were not quickly absorbed into the work force.

One problem was that many skilled workers were capable and experienced in only a portion of the broader craft. Thus skilled workers in railroad shops became so specialized in their duties that they were unable to follow their craft in other industries.[21] Older printers found it difficult to adapt to the regimen and techniques needed to operate the Linotype. Thus they were forced from their jobs, and it proved difficult for them to enter a new trade.[22] In the mid-1890's, the Massachusetts Board to Investigate the Subject of the Unemployed examined the shoe industry in the state, and it concluded "some of the labor displaced has been re-absorbed by increased production, but not all; and the ratio of unemployed slowly but steadily increases."[23] In the metal trades, employers reported that increases in production had reemployed men displaced by machinery; but the board found in the copper industry that although the total number of men employed was the same, unskilled men had replaced the skilled during the preceding decade.[24] In the morocco leather industry of Woburn, from 25 to 40 percent of the skilled men had been displaced by machinery, and while some had been reabsorbed because of increased demand for the finished product, "unquestionably many are permanently out of work."[25]

Certainly the most widely discussed instance of technological displacement in this period was the impact of the Linotype on printers in the 1890's. Although the depression of 1893 led to increased unemployment throughout the economy, observers of the printing trade laid the hard times of the mid-1890's to the introduction of the Linotype. They considered it a case of technological displacement, not cyclical unemployment.

The Maryland Bureau of Labor Statistics reported in 1894 that "the introduction of typesetting machines has been very largely instrumental in displacing hand typesetters in Baltimore, as elsewhere. . . ." Some of those displaced found work in the Government Printing Office in nearby Washington, D. C., where hand methods continued.[26] In 1896, the bureau noted that fifty-one Linotype machines had been introduced in Baltimore since 1891, and each machine displaced three men. This figure would have been greater except that some hand compositors had to be kept in newspaper offices to set display heads and advertisements that were not done by machine.[27]

The Massachusetts Board to Investigate the Subject of the Unemployed reported that the introduction of machinery had reduced the printing staff of a large Boston newspaper from 200 to 120 persons. Overall, about 10 percent of the regularly employed printers in the state had been displaced by the Linotype. The figure is low, as compared with other estimates, because book and job printing were included, and machinery had no impact in these areas of the trade. Many of those displaced had gone to the South or West to seek jobs as hand compositors in areas where machines had not been introduced. Two-thirds of the compositors displaced were unemployed and "still hoping for the re-establishment of the old system."[28]

In St. Louis, approximately one-half of the hand compositors had been displaced, while estimates in St. Joseph, Missouri, set the figure there as anywhere from one-third to two-thirds of the compositors. Of these, a good number found employment in job printing, where typesetting had not been mechanized, or in setting display work by hand. However, many had been forced from the trade.[29] Local 6 of the International Typographical Union provided figures on displacement for New York City. By 1895, 38 percent of the regular printers and 39 percent of the substitute men had been displaced in the nineteen composing rooms included in the survey.[30]

The experience of the printers was widely discussed as evidence of the displacing effect of machines and as concrete evidence that unrestrained innovation was disastrous to the worker. However, the long-term situation was hardly so bleak. Since the Linotype was best operated by a skilled worker, hand compositors generally took the jobs on the machine. This immediately reemployed many of the compositors at wages comparable to those earned at hand work. Wages subsequently increased as the men became expert in the use of the machine.[31] Moreover, the Linotype cut the costs of production sharply, and its use led to a rise in profits for the employer as well as strong competition among publishers. The result was an increase in both the size of newspapers and the number of editions, which greatly enlarged the employment for hand compositors who could learn to operate the machines efficiently. Even in the years before 1897, changes in the operation of the composing rooms of newspapers, in response to the flexibility provided by the machine, increased the demand for additional printers and reduced the initial displacing effect of the Linotype.[32]

In addition, the transition from hand to machine labor in printing was accompanied by a reduction in hours from an average of ten to eight per day, mostly in the years after 1897. At the same time, the productivity of the printer increased at least fourfold. Although the reduction in hours did not equal the increase in productivity, it clearly helped reduce the burden of displacement in the trade.

Despite the relatively successful experience of the printers, it could not become a model for all workers facing technological change. Not all trades had so strong a union, one that could negotiate for the employment of its own members as machine operators. Not all trades faced innovation that still required considerable expertise and made it worthwhile for employers to continue to use skilled men in preference to more poorly paid semiskilled or unskilled workers. Thus the Linotype did not eliminate skill, it merely made the skilled workmen immensely more productive. In such a situation, it was to the interest of employers to have a well-trained labor force. Thus the union was able to take over the new jobs for its members, reduce hours, maintain or increase wages, and retain its strength—all with a significant decrease in costs for employers, and with benefits to the consumer in the form of cheaper newspapers.

Even within this highly favorable situation, significant displacement did take place in the years immediately following introduction of the Linotype. Although employment expanded as demand increased and hours were reduced after 1897, many workers undoubtedly left the trade before that time. Labor leaders always stressed that an ultimate gain in employment, because of machinery, did not solve the immediate problem of displacement, unemployment, and poverty for workers who lost their jobs for some undetermined period. Also, many innovations did not increase the need for the same type of worker. The number of jobs might ultimately increase in an industry, but unskilled or semiskilled workers would fill these places rather than the older skilled labor force. In fact, the introduction of machinery could be the means by which an employer would rid himself of the skilled worker and his union.[33] Thus the experience of the printers did not invalidate the fears among skilled workmen that mechanization would have a destructive effect upon their jobs and lives.

Displacement of workers in any trade had effects in other trades. Although many crafts were affected by innovation at the same time, the rate of displacement and the degree to which skill was diluted varied. Thus men forced from one craft often sought work in other trades that had also experienced a dilution of skill but which still offered work at wages above those to be earned as machine tenders in their original occupation. The carpenters complained that innovation had so simplified their work that workers displaced from other occupations easily learned enough carpentry to enter their trade, driving down wages and creating a constant surplus of labor.[34] Ironically, the carpenters were also displaced, and their sons sought out trades less affected by mechanization, such as plastering.[35] Formation of the International Association of Machinists in 1891 was stimulated by the entrance of poorly trained men from other trades. To meet this problem, unions in all trades were urged to oppose interlopers.[36] Apprenticeship regulations, the refusal of union men to work with nonunion employees, and demands for state licensing of certain craftsmen, to insure a minimum level of skill, were the major devices used by unions to seek to limit entry into their particular trade.

The labor movement also recognized that technology created or intensified seasonal unemployment. From the early days of the trade union movement, labor leaders had linked seasonal unemployment to the division of work, since a given amount of goods could be completed in a shorter period. This development was considered a major threat to the interests of the worker. Machinery only intensified this situation by speeding up production even more. Thus a smaller force of men could fill the needs of the market in less time, displacing some and creating longer periods of idleness for those who remained.[37] Seasonality also undercut the bargaining position of a worker in several ways, the most important being its effect upon savings, and thus the worker's ability to sustain a strike once the season resumed. It also added to the pressure on workers who were employed, as idle laborers sought work elsewhere during their own dull season.

In response, trade unions turned to tactics that would stretch out the work. Various work rules were adopted in order to restrict output.[38] These rules had a number of objectives, but a principal one was reducing seasonality to a minimum by offsetting the tendency of industrial innovation to complete any job, or meet any demand, more quickly.

Defenders of innovation argued that improved methods of production would increase demand by reducing costs, which ultimately would require a full year's work based on the expanded market. Labor leaders were unsympathetic to this argument, because their experience indicated that seasonality continued, even in trades long mechanized, such as the making of shoes, hats, and textiles. The worker was asked to accept a present evil for a future gain; but since he had little control over the direction that the market would take, and since even employers seemed unable to predict the course of demand or plan for it efficiently, such an argument seemed to be little more than a rationale to justify gains for the employer as against his employees. As we shall see, one of the constant themes during America's industrial revolution was a supposed overproduction, which only fortified the resolve of trade unions to regulate the productive process if they could. In an economic situation of competitive self-interest it was a rational choice.

The shorter work day was also a device for offsetting the tendency toward increased seasonality. Robert Schilling, President of the Coopers' International Union, argued in the 1870's that a shorter work week would mean "not so much work will be made and where now men work probably only half the year, the work will be distributed more evenly and last longer, while better prices can be obtained."[39] Jonathan Fincher also viewed shorter hours as a necessary step against the tendency of the seasonal layoff to rob workers of their independence by undercutting their bargaining position. However, Fincher was perceptive enough to realize that eight hours was far from a complete solution, and that it might actually create additional problems for workers. Although shorter hours would tend to lengthen the busy season, it was not likely—even before mechanization so speeded up the productive capacities of industry—to eliminate the dull ones. Thus "in times of slackness, the old rule of discharging a portion of the men employed, and retaining others at reduced wages, would work the same demoralizing effect; but perhaps to a lesser extent." Moreover, shorter hours would increase the number of men employed in the trade, and unless seasonality were eliminated, the number of unemployed men in the dull period would be larger, intensifying the evils already resulting from such idleness.[40]

While Americans were constantly reminded of the benefits conferred by industrialization, there arose a real concern, in the last two decades of the century, about a supposed increase in crime, pauperism and tramping that seemed to be another feature of economic change. Much of the blame was laid upon the new immigrants from Southern and Eastern Europe, and the fear that they were a criminal (as well as a radical) force helped fuel the intensive campaign for restriction of immigration in the 1880's.[41] Labor spokesmen realized the public's concern for safety and order, and thus there were repeated efforts to connect technological unemployment to these social ills.

The Machinery Constructors' Association of North America put the case most concisely. "There is a steady increase of labor saving (or idleness creating) machinery. There are more men than work. Statistics demonstrate the fact, crime increases because idleness increases. Idleness and crime are twin scourges, they are in intimate companionship."[42] In a more vivid but equally direct fashion, R. S. Lampton lamented a poor worker's loss of his job to the labor-saving machine. Although the workingman sought another job, he found none, and so sent his wife to beg on the streets. Ultimately, he decided on his only course of action.

> I said good-bye to my darkened home,
> The scene was black despair;
> The children sang a Sunday School song,
> And Mary offered a prayer.
> Only the home of a workingman,
> Dark and dismal and damp;
> The labor machine has got my job,
> And, brothers, I'm a tramp![43]

The immigration issue deeply divided the labor movement. The carpenters' union was generally a force for restriction, in large measure because of the competition of European workmen in this trade. However, the Albany Local 274 of the Brotherhood of Carpenters and Joiners tried to argue against laying the blame for unemployment on immigrants. "Machinery is the only competition that counts. Machinery rents no houses, consumes no groceries, gives back nothing to employ others like the living, breathing employee. It has come to stay and force men to idleness and public support in public institutions—charities and crime."[44]

The solution to these problems lay in relieving idleness through the universal use of the eight-hour day. This would convert the tramp, and other members of the "dangerous classes," into consumers and producers and alleviate many of the problems that such persons posed to the society.[45] This line of argument aimed to link labor's interests to the concern of a broader constituency. As we shall see, a more significant attempt to achieve a similar end was the argument that unrestrained mechanization generated overproduction, which meant economic imbalance and ultimately crises that injured the interests of all members of society, not only those of skilled workmen.[46]

Most commentators on mechanization agreed that one group of workers was particularly hard-hit by innovation—the older skilled worker. Such workers were regarded as less adaptable, less able to keep up with the pace of the machines.[47] Also, younger workmen had invested less time in acquiring whatever skill they had and thus suffered less of a sense of loss in becoming machine operators.

In a preindustrial system, older workers were respected because they had a real and continuing value in the productive process. Skill and experience were attributes of age that were only partially reduced by physical changes in the advancing years. The machine destroyed the economic value of older craftsmen by substituting energy for skill, and thus created the problem of what to do with men who were now not only useless in their trade, but likely to become a burden to those around them. Old age had always been a time of trouble for the unskilled, for as physical energy failed, so did a man's ability to earn his way. One of the incalculable appeals of skill had been the productive role that older men could continue to play. To take this from men who had expected to be respected workers through much of their later life was to reduce such persons psychologically as well as economically.

By the last decade of the nineteenth century, tentative proposals were made to meet this problem. One approach was adopted by the International Typographical Union, which established a home for aged union members so that printers displaced by the Linotype would not face the debasing alternatives of dependence upon family, resort to charity, or loneliness and steady decline in poverty.[48] The hatters' union was also urged to establish a national home for its aged members.[49] More important to the majority of older workers, who did not wish to live apart from their family and friends, was a system of old-age pensions to meet the economic losses of displacement, if not the emotional ones.[50] However, few pension plans of any sort were established during this period.

In the last decade of the nineteenth century, reformers increasingly directed their attention to the problem of unemployment. However, their interest in the subject did not lead to cooperation with organized labor because of fundamental differences in approach and thus in the specific solutions that were offered. Reformers did accept organized labor's basic refusal to support unrestrained industrial change. However, instead of relying on the actions of trade unions as the basic mechanism for control, the reformers turned to substantial and continuous action by the state.[51] Industrial growth would continue through private initiative, but the state would cushion its effects upon workers, and society in general, by legislation aimed at specific evils. Although the state's action at any moment was specific, it was understood that there was no single solution for industrial problems. The state would have to act continuously to meet the most serious difficulties. At this time, reformers rejected restraint or control of the basic economic decisions made by individuals and corporations, but even continuous state action to meet the results of such decisions was a fundamental reordering of American values. As one reformer put it in 1887 in discussing the more equitable distribution of income that would accompany reform: "There are those who seem to think that whoever finds fault with the present system of distribution is demanding that the rich shall be robbed for the benefit of the poor, whereas, the fact is that they are simply insisting that the poor shall not be robbed for the benefit of the rich."[52]

Discussion began among reformers about the feasibility of some system of compensation to meet the destructive effects of idleness. At a later point, the leading reform economist, John Commons, would insist that unemployment insurance could be a force to discourage idleness as well as to alleviate its effects; but in the first instance the emphasis was upon using compensation to meet the human and social problems that flowed from unemployment.[53]

An immediate problem in discussing any form of unemployment insurance was to determine eligibility. By 1900, there was general agreement among reformers that the able-bodied unemployed could be divided into "those who *want* work and those who *won't* work." Those who could not work were a separate problem that could best be met by more traditional relief through charitable sources. However, society had a real responsibility to those who wanted work but were idle, since the economic system was responsible for such unemployment. Failure to act was disastrous for workers, who depended upon steady wages for their livelihood, but also dangerous to society, since "Revolutions are born of neglected wrongs."[54]

If unemployment was an inevitable result of an industrial economy, George Gunton believed it was prudent to provide an adequate insurance system in order to undercut opposition to technological advances. Gunton was convinced that mechanization would ultimately benefit the displaced workers along with the rest of society. The problem was to tide the worker over the period of transition until he was able to find new employment and thus enjoy the benefits of a mass production economy.[55]

Employment was obviously so crucial to the worker that it could be viewed in the same light as property. The state compensated those who lost property through eminent domain; why should it not compensate those who lost employment? Workers had a right to work—a new right created by the growth of an industrial economy. "The laborer has been encouraged by society to fit himself for a particular trade, and when this trade is abolished in the interests of society, the employer, first, and society, ultimately, should share the loss with him."[56] Professor Charles Tuttle argued that the money to do this should be raised from those who secured the benefits of mass production. Thus employers who have gained larger profits, and the general community, which has secured a greater volume of goods at lower prices, should be taxed to indemnify those who had suffered from industrial change. The unemployment benefits should be large enough to allow the worker time to find another job.[57] Clearly this was an effort to achieve both technological advance and economic justice.

The proposal for unemployment compensation linked together displacement (which, in effect, was similar to the loss of property, since the skill so essential to the artisan's particular position in society was permanently removed), recurrent idleness of a seasonal and cyclical character, and unemployment produced by the failing of a single company or the particular conditions in an industry. The reformers focused on the most dramatic form—displacement—although their remedy was actually better suited to the type of idleness that did not involve loss of skill.

Proposals for unemployment compensation received a cool reception among conservatives and many reformers as well. Professor William Folwell regarded Tuttle's proposal as wholly impractical. Did Tuttle propose "to raise a fund in advance, to be kept in some treasury ready for use when some industry, at some unknown and unexpected time, shall be destroyed by a new process or a new invention? We don't do things that way in the United States."[58] George Roberts, the Director of the United States Mint, wondered why workers should be compensated for loss when employers who suffered from technological advance received no assistance.[59] The prominent charity worker, editor, and reformer, Edward T. Devine, believed an unemployment indemnity would "put a bonus upon inefficiency," and he argued that relief during periods of idleness should be treated as a form of charity, "with direct reference to the actual experience of the one who is displaced."[60] Devine thus rejected the concept of compensation as a right of the displaced worker, and instead offered assistance as a privilege, subject to the restraints and conditions imposed by the charitable organization. Charity workers had traditionally regarded such controls as essential if there was to be rehabilitation as well as relief, and Devine was prepared to apply the same principle to unemployment.

Perhaps the most significant objection to unemployment compensation came from persons who were prepared to accept the justice and necessity of some form of unemployment relief, but foresaw serious difficulties in administering such a plan. William Willoughby wrote America's first comprehensive treatment of social insurance in 1898, and he supported workmen's compensation, sickness insurance, and old age pensions. However, he was wary of unemployment insurance, since it was so much more difficult to calculate risk in advance and adjust contributions to the degree of risk to keep the insurance system solvent. Also it was much more difficult to determine if an individual had a valid claim. There were few reliable tests to determine whether a man had conscientiously sought work, nor were there criteria concerning the conditions under which he might refuse employment and still be considered unemployed.

Willoughby believed that these problems precluded a state-operated system of unemployment insurance, at that time, and he suggested that any further developments in the field should utilize trade unions, since they were best able to determine whether an individual was truly eligible.[61] Several European states adopted such an approach in the first decade of the twentieth century, and this so-called Ghent Plan continued to draw serious attention in the United States until after Great Britain introduced a state system of unemployment insurance in 1911. American reformers did not develop a well-formulated plan for an unemployment system until just before World War I, and though John Commons lobbied staunchly for his own version of unemployment compensation in the 1920's, the enactment of such legislation awaited the Great Depression and its catastrophic effects upon employment. At that point, technological factors, which had been so vital in the original discussion of unemployment compensation, were less important than the general failure of the economy.

Another major source of opposition came from the labor movement itself. Labor leaders opposed unemployment compensation on particular grounds and as part of a general hostility to continuous and substantial action by the state.

Supporters of unemployment compensation refused to accept the basic contention of labor leaders that mechanization produced poorer jobs for those displaced by destroying skill, which thereby created conditions where the new supply-and-demand relationship worked more strongly in the interests of the employer. Reformers shared the general view that real wages would increase over the long run because a possible loss in money wages, through reemployment as a machine tender, would be compensated for by the lower prices made possible by the same machine. Moreover, reformers were less willing to accept the widespread belief among trade unionists that wages would not necessarily rise with gains in productivity. Thus unemployment compensation met the problem of the transitional period when skilled workers were without work. The reformers concentrated on the quantity more than the quality of reemployment, and they regarded it as society's obligation to compensate a worker for the loss of employment, not for the degradation of his economic or social position.

Beyond the issues involved with unemployment compensation, all forms of state social insurance received little support from the craft unions that dominated the American Federation of Labor. Although some labor leaders accepted the need for labor legislation and a continuous government presence, they were mainly from unions that were not strongly entrenched in their industry and that seemed unlikely to become so. Thus law might accomplish what collective bargaining could not. However, trade unions that exerted real influence over conditions in their industry were reluctant to support legislation for fear that the state ultimately would replace the labor organization as the protector of the workers' interests. This was unacceptable on several grounds. First, it might slow the growth of trade unions as workers turned to the state for assistance; but, more important, it threatened to make the conditions of employment dependent upon an institution beyond the control of labor. The state had traditionally been more eager to

serve the interests of the rich and powerful. To place the interests of workers in the hands of the state, without political power sufficient to insure
that government would act in the interests of labor, seemed dangerous.[62]

Samuel Gompers's doctrine of voluntarism arose from these considerations, and it represented the views of the stronger trade unions. Gompers
believed that workers had to improve their situation through the trade
union because this was the only organization under the control of workers.
In contrast, he opposed action through the state because of the political
weakness of organized labor. Political power for labor had been elusive in
America, and seemed no nearer at the end of the nineteenth century. Gompers also discounted the possibility of a broad alliance of workers, elements
of the middle class, and reform-minded intellectuals. He stressed class barriers based upon differences in economic function and interest. Only when
the economic interests of social groups roughly agreed was cooperation
possible, and Gompers believed that labor's objectives were sufficiently dis-
similar from those of any other group to insure that an alliance would ultimately founder. Only organizations of workers could be counted upon to
defend the interests of workers.[63]

Unfortunately, little was done during the American industrial revolution
to meet the problem generated by unemployment. Seasonal and cyclical
layoffs affected the skilled and unskilled alike, and those out of work were
left to their own fragile resources or to the inadequate ministrations of the
charities. The displaced artisan lost more than income; he gave up the skill
that had been the buttress of his social position in society. Loss of skill
clearly was linked to loss of status as skilled workers increasingly were trans-
formed from artisans, who labored in small shops and aspired to become
owners of their own small enterprises, to operatives in factories. The American mobility ethic had to be reexamined in terms of the new conditions
produced by this basic change.

## CHAPTER THREE

# INDUSTRIALIZATION AND THE STATUS OF WORKERS

American mythology gave great value to independence and entrepreneurship, which were linked together in the figures of the pioneer and the self-sufficient farmer. The independent master craftsman who owned his small business was part of the same image. These activities not only carried positive status but in the early nineteenth century were still achievable (though not necessarily achieved) by a significant portion of the population. However, the skilled journeyman rapidly lost this element of possibility. Even in the first half of the nineteenth century, many journeymen remained workers all their lives.[1] In the 1850's and 1860's, the increasing division of labor further reduced the role of skill as an entry into the entrepreneurial ranks. Some labor leaders hoped that this situation could be reversed by a revitalization of skilled labor, but this possibility became increasingly unlikely in the face of the substantial amount of capital required by division of labor, mechanization, and factories. The need for capital destroyed the traditional basis for entrepreneurship, which had rested on skill and experience. However, the traditional mobility ethic continued to exercise much influence over popular attitudes throughout the last half of the nineteenth century.[2]

To the extent that workers continued to believe in the mobility ethic of a preindustrial age, they were ill-equipped to meet the rapidly changing conditions of industrialization. This basic point underlay the appeal of the New York Wood Carvers' Association to skilled furniture workers in 1874.

The producing method peculiar to capital is that, starting from unlimited competition it must conquer the market by cheapness, and therefore constantly produce in large quantities, which again can only be done by the minutest division of labor. Ingenious machinery has then to be used to such a degree that skilled workingmen—with the exception of machine builders—become superfluous. The inevitable consequence of such a system is that all mechanics must become factory *workmen,* and all production machine *work.* There is no escape from that.

If this were true, workers had to realize that individual action could not solve what were essentially class problems. In terms reminiscent of William Sylvis, the workers were urged to recognize that modern industry had created an unbridgeable gap of interest between worker and employer, and that unionization was thus necessary to defend the real interests of workers. It was also crucial to recognize that such a union would not try to block "social progress," but would attempt to combat the injurious effects of industrial change.[3]

In essence, this represented the thinking of most trade union leaders from the 1860's through the rest of the century. The labor movement that revived after the depression of the 1870's clearly stressed the need for class advancement through trade unions. The older tradition of individual mobility insured that the many would gain little despite the possible successes of the few. In the mid-1880's, Thomas C. Weeks, Chief of the Maryland Bureau of Labor Statistics, believed that industrialization was rapidly providing the conditions that would create such unions. The end of the realistic hope for mobility into the entrepreneurial class, the destruction of skill, and the definition of labor as machine tending destroyed ambition, interest in one's work, and the sense of independence. It left the worker with a sense of helplessness, and this inevitably led to the formation of trade unions.[4]

By the 1880's, social critics clearly recognized the link between industrialization and the end of older concepts of mobility. Henry George argued that the machine was the root cause of the concentration of wealth and economic power in the hands of the few. It fostered the division of labor and large-scale production, which ultimately destroyed skill and the independence of the worker. The mobility that once existed between skilled labor and the employing class was being rapidly destroyed. George believed it was a gross distortion to continue talk of mobility as if the United States were still a nation taking possession of an untapped continent. Mechanization and the concentration of power in the hands of the wealthy capitalists had closed opportunities once generally available. "When a railroad train is slowly moving off, a single step may put one on it. But in a few minutes those who have not taken that step may run themselves out of breath in the hopeless endeavor to overtake the train." Some few would still advance, but the mass of workers were dependent machine tenders who would remain such. Small businessmen would be driven from the field by larger concerns.

It was a process akin to European feudalism in result: the subordination of small property owners to the wealthy and the ultimate domination of society by the few.[5]

Richard C. Ely was a leader in the attack on traditional economic and social thought during the 1880's. He believed ninety-nine out of every hundred workers would remain in the laboring ranks, since industrial change had replaced the small shop with the factory. Not only was mobility largely blocked, but the small producer was rapidly losing his independent position and falling into the wage-earning class. The mobility ethic was thus fallacious and dangerous, since it offered as reality a social mechanism that had become anachronistic. Instead society should seek "to improve the laboring man as a laboring man—for such the great mass must remain for many years to come. . . ."[6] Although Ely and George differed on the method for such improvement, they agreed that refusal to recognize the significance of the end of the mobility ethic was at the very root of the "labor question" that so troubled America in the 1880's.

Those who believed in some variety of cooperative ownership of the means of production largely accepted the picture of society painted by George and Ely. Peter J. McGuire, a socialist leader in the 1870's and General Secretary of the newly reconstituted carpenters' union, argued in 1883 that there was no possibility for increased mobility through workers becoming small shop owners. To accomplish this would mean smashing machinery, an impossible course, and one that was undesirable since industrialization offered much to the worker once the existing system had been replaced with cooperative ownership.[7]

Charles Litchman of the Knights of Labor also stressed the degradation of the worker who lost his skill to "become part of a machine"; but like most critics of mechanization, Litchman made it clear that "No sensible man would turn back the hands upon the dial of human progress by abolishing machinery." Workers must achieve ownership of machinery through producers' cooperatives, which would eliminate the evils of machinery and allow the benefits to be shared by all.[8]

Litchman regarded his particular form of cooperative ownership as an ultimate successor to the wage system and the established mobility ethic, but workers could join producers' cooperatives in an attempt to retain their mobility within the existing capitalist system. The producers' cooperative could be a device for securing from many workmen the capital beyond the means of any one of them, which would allow them to participate as members of the entrepreneurial class. It could be an attempt to retain some vestige of the old mobility ethic, not a step toward social reconstruction.[9] Unfortunately for these cooperators, industrial giants could crush producers' cooperatives as easily as small, individually owned businesses, and combined with a host of internal problems, such competition destroyed most of the cooperative shops.[10]

Closely related to the end of mobility from the skilled crafts into the entrepreneurial class was the destruction of the existing small businesses and the concentration of capital in the hands of a few large corporations. This development led to discussion of whether the interests of the worker were served by the destruction of small business. Small employers argued their inability to meet the lower prices that resulted from the use of machinery in factories. Consumers were attracted by lower prices, even if they were accompanied by lesser quality, and the small employer, using skilled labor and a minimum of machinery, often could not compete once the railroads opened their local markets to mass-produced items.[11] Some argued that this situation hurt the skilled worker as much as the small businessman. A manufacturer of tinware in Kansas reported that "the large tinware manufacturer can produce the ware cheaper than the wages we would pay to regular tinners for making the ware, without counting the material that it is made of." Thus he had been forced to take on odd jobs, and the result was an attrition in his work force, in ten years, from ten or fifteen full-time hands to three.[12] The repetition of this situation in trade after trade meant that employment would eventually be concentrated in relatively few companies.

To some, this was a dangerous situation, since jobs would be more secure if distributed over many small firms than concentrated in a few large concerns.[13] However, influential elements within the labor movement were unafraid of the large company. President John F. Tobin of the Boot and Shoe Workers understood that the introduction of machinery in a competitive market situation placed extreme pressure upon the smaller manufacturer, who lacked the means to purchase the newest machines and thus keep his costs down. Instead, the small manufacturer tried to cut wages, producing constant conflict with the union.[14]

An industry dominated by small concerns often was marked by vigorous price competition, which made an employer reluctant to raise wages unless his competitor did so as well. Unless markets were local, such a situation made it necessary for the union to organize effectively throughout the industry—which often proved difficult for the young labor organizations of the period. When large concerns controlled a good portion of the market, they were often able to pass on wage increases to the consumer. Labor leaders such as Samuel Gompers, long-time President of the American Federation of Labor, believed that under such conditions the large company might view collective bargaining and the resulting labor stability as being in its own interests. Accordingly, Gompers joined the National Civic Federation, which included some of the nation's most prominent financiers and businessmen.

Little advantage came to organized labor from Gompers's participation in the National Civic Federation, but clearly it was an indication that many trade unions were quite prepared—and even anxious—to welcome the large corporation if it would accept collective bargaining.[15]

In addition, the socialists within the labor movement accepted the trust as a necessary step in the concentration of capital and the sharpening of the class struggle. Socialists opposed any defense of small business, because it tended toward the concept of a "producing class" of employers and workers who were allied against the plutocrats. This not only enlisted workers in the defense of the propertied interests of the middle class, but also blunted the necessary development of the workers' sense of class consciousness.

To the extent that the mobility of skilled workers was restricted by industrialization, so was their status in society downgraded. Intelligent conservatives recognized that this undermined one of the traditional supports of American capitalism. Edward Young, Chief of the United States Bureau of Statistics, pointed out "the journeyman mechanic who could see a prospect that within ten years he might himself become the owner of a shop was not disposed to feel or act unkindly toward a class of which he so soon hoped to become a member, viz. the employers; but to the operative the possession of a factory is a thing so remote from probability that it scarcely enters into his wildest dreams of future success."[16] Thus the factory system had the potential for dangerous social consequences, for to the degree that a permanent industrial proletariat was created, social unrest was likely to result.

One possible solution to this problem was to dislodge status from mobility and place it in manual labor as such. Thomas Bigham, Commissioner of Statistics for Pennsylvania, commented that "the little one-horse work-shop of the past is too slow for the seething, surging activities of the present . . . and the place of the little shop at the corner of the street or the crossroads, is occupied by the grand manufactory, glowing with the flame of its hundred forges and roaring with the flash and turmoil of its thousand workmen. Now, does he, who today in that grand factory, working for wages, represents the small operator of yesterday in the two-benched or one-forged shop at the crossroads, change his status by becoming, in the new order of things, a wage worker?"[17] Despite Bigham's enthusiasm for the new industrial setting, a worker, with little hope of mobility out of the wage-earning class, did lose status. In the 1880's, machinist John Morrison characterized the situation in his trade as a drop from marginal middle class to lower class.[18]

In 1884, Edward T. Peters of the United States Department of Agriculture brought the economic and social effects of machinery to the attention of the American Association for the Advancement of Science. He stressed that the wage earner of the 1880's could no longer be compared with the worker of a half century earlier. In a preindustrial economy, workingmen owned much of the industrial capital through the small businesses that dominated the various trades. In the 1880's, the worker was totally dependent upon a separate capitalist class, and any discussion of the "labor question" that focused only on the improvements in the general standard of living in the nation, without reference to the conversion of artisans into simple wage earners, was useless. The dependence of workers upon a separate class of capitalists was a result of the destruction of skill by machinery, and while Peters was not prepared to forego the advantages of mechanical progress, he believed "such dependence is unfavorable to the highest type of manhood. . . . The manhood of a nation is its most precious possession, for the loss or deterioration of which no increase of material wealth can adequately compensate." Peters understood that the old system of small shops was gone, but the workmen must still be given a share in the capital of the nation through "the ownership of a proportional share of stock in some larger establishment." He mentioned the cooperative building societies as one possible vehicle for the basic task—the reuniting of capital and labor in the same hands. Such a step was necessary to prevent the development of a disaffected working class that might seek to remedy existing ills by "an erroneous method of treatment."[19]

This dependency theme, with its resultant loss of status for the worker, was an important element in the social thought of the period.[20] The threat to the stability of America's society that was inherent in the collapse of the older values demanded a new system that could continue to bind workers to the existing economic arrangements. The solution proved to be the substitution of property and possessions for mobility and independence as esteemed goals and marks of success. Conservatives accordingly stressed that industrialization cheapened the cost of many items once available only to the rich, increased real wages, and thus allowed the worker to enjoy a higher standard of living than under any hand labor system. The major features of the "good old days" had been the status accorded skilled workers and the possibilities for mobility, not material goods. The industrial revolution destroyed these older elements, but allowed workers new measures of success and acceptance centered around material acquisition. There was more stress on home ownership by workers. Progress for the mass of Americans increasingly was measured by the ability of average citizens to afford possessions once available only to the rich. The image of the anxious consumer replaced that of the independent artisan.

These criteria for determining success were in better balance with the realities of an industrializing society, and American workers, both immigrant and native born, accepted the newer measures of status.[21] In a period of mass production and economic growth, possessions were a plausible, though elusive, goal; but once achieved, material acquisitions earned regard from a middle class that prized the same goods. The new standards for measuring success and status clearly satisfied Bigham's wish by allowing for the conversion of artisans into industrial workers while creating social values that would continue to bind them to the existing economic order.

Home ownership, for example, helped tie the worker to the existing system through the meshing of tangible property and intangible, but real, achieved status. As one observer noted in 1879, once established in even so modest a place within the property-owning classes, workers would not be likely to challenge those owning more property for fear it would disturb their own holdings.[22] In addition, workmen would be more docile, since they would fear dismissal all the more for its effects upon their ability to maintain payments on the mortgage. The emphasis on possessions and property also had another substantial benefit for American businessmen. Accumulation not only became a mark of success, but it had the indispensable function of creating markets for the flood of goods unleashed by industrialization. Thus the newer measures of success served the interests of America's propertied classes, but also had sufficient plausibility and appeal to win the sufficiency of approval from wage earners necessary for any status symbol to be effective.

As the possibilities for advance into the entrepreneurial ranks declined, the redefinition of the criteria for positive status placed an even greater premium on the standard of living, which allowed the worker to secure the possessions and follow the mores of the middle class. Some artisans had been able to improve their condition, educate their children, and occupy more comfortable homes. These workers were restless as machinery threatened to force down their standard of living to the level of the unskilled.[23] To the degree that such workers had assumed the trappings of middle-class existence, they received a social regard not given to the poor in the labor force. Thus a reduction of income would also mean a decrease in status.

Whatever the ethnic, religious, or racial character of the unskilled, they were usually poor, and poverty has always been despised in America. Society has denied responsibility for idleness, crime, drunkenness, overcrowding, disease, and a host of other social ills. Instead it has shifted the blame to the poor themselves. Accordingly, social status has been tied to the distance from the mores and behavior of the poor, and the closeness to the behavior patterns of the middle class. For the artisan, his skill was thus a means of escaping opprobrium by earning enough to live more like the comfortable middle class and less like the poor. Thus the unskilled not only threatened the wage level of the skilled worker, but, in the process, the craftsman lost the respect desired from other elements within the society.

In addition, differences in status were paralleled by the competition between the skilled and unskilled as mechanization increasingly opened trade after trade to the poorer, untrained worker. Writing in the 1880's, Terence V. Powderly, Grand Master Workman of the Knights of Labor, claimed that the exclusiveness of the trade union had been the basic reason for the formation of a broadly based labor organization such as his own. Skilled workers objected to associating with the common laborer, for "soon they will take our places at the bench, and it is time to nip this thing in the bud."[24] Although the apprenticeship rules of trade unions were designed to prevent the unskilled from replacing an artisan, Powderly claimed that the skilled had few compunctions about taking the place of the poor laborer should that be necessary. Thus, "The rights of the common every-day laborer were to be considered by the new order [Knights of Labor] because the members of trades unions had failed to see that they had rights."[25]

Powderly wrote this in the midst of his conflict with the American Federation of Labor, and he ignored the fact that many of the early units of the Knights of Labor were also composed of craftsmen; however, the hostility between skilled and unskilled seems clear. The leading labor journalist, John Swinton, wrote consistently against a fracturing of the working class into skilled versus unskilled, which could only serve the interests of the employers. Moreover, trade unions made no effort to hide the threat they perceived from the competition of the unskilled worker, increasingly made more dangerous by immigration and mechanization.

As early as the 1870's, the furniture workers faced the elimination of skill through mechanization. The result was extensive use of unskilled labor, often women and children, and a drop in wages, "in some cases below that of the common day laborer." Thus although employment in the trade did not decrease, the conditions of labor worsened.[26] The long-time labor leader and reformer George McNeill argued that lower wages were inevitable when workers were "compelled to work on the basis of payment for time of service rather than . . . skill."[27]

In the same period, carpenters were faced with the manufacture of doors, sashes, and the like by machinery in factories. The carpenters no longer made these items on the job, but instead installed the ready-made goods. As a result, many more men were able to acquire this diluted skill.[28] Carpenters regarded their lower wages, relative to other crafts in the building trades, as a direct result of the conversion of skilled jobs into semiskilled ones.

The same problem later invaded other building crafts. Plumbers attacked the cutting of pipe by machinery and the building of tubs and sinks with couplings so simple a "laboring man" could put them together. Even the skill needed for the jointing of pipes was endangered by the star joint, with the result that wages would drop from $3 or $4 a day to the $1 or $1.50

earned by the "men of the streets" who installed gas and steam pipes in the public roads. The response to these dangers included pressure for local building codes that would maximize the need for plumbers, and demands by the union that all pipe cutting machines be operated by journeymen plumbers at union rates.[29]

The painters similarly faced a threat from ready-mixed paints. As was true in many crafts, the innovation was attacked as inferior to the hand-made product, and thus as a liability for the consumer as well as the workmen.[30] Painters regarded the ready-mixed product as an inducement for incompetent workers to enter the craft "outside of regular trade channels."[31]

The skilled garment cutter objected to the long knife, an improved tool which greatly decreased the strength and skill needed to cut cloth. Its introduction opened the craft to many more workers. The United Garment Workers condemned the long knife because it reduced skill and thereby allowed employers to "demoralize the craft." A New York local vainly set up a special fund to curtail the use of the long knife.[32]

By the end of the nineteenth century there were few trades which had not been affected by mechanization, yet the continued improvement of machinery periodically renewed the threat to the existing level of skill. The hat industry had abandoned many hand processes by the mid-nineteenth century. However, in the late 1880's and 1890's, new machinery threatened to eliminate still further the element of skill. Martin Lawlor, a leading union official, claimed that such machines had eliminated hand sizing and created a surplus of journeymen. The result was a decline in wages as the workmen idled by mechanization competed with less-skilled newcomers for jobs as machine operators.[33]

These examples might be multiplied many times, but they clearly reveal the capacity of mechanization to divide workmen along lines of competing self-interest. The skilled viewed the unskilled as competitors. In the capitalistic system of the period, no restraint on such competition existed except what the skilled workers might devise through their trade unions. It was hollow to preach to skilled workers about the selfishness of such defenses when no viable alternative seemed to be available, and when the employers used the unskilled to depress wages in the industry. The advantages of innovation for the public were often compatible with profit; but skilled workmen felt that progress should not be defined as cheaper prices for the consumer and greater profits for the employer but lower wages, increased idleness, and reduced status for the producer.

In the twentieth century, government increasingly has modified unrestrained industrial change. This is not to say that the interests of workers are necessarily protected by government, only that they may be. The actual result is part of the same political process that determines so many other basic elements of American society. Trade unions continue to exert independent economic pressure on behalf of their members; employers continue to use mechanization to increase their profits, often at the expense of their workers' interests; but the government now serves as a third party which can work out programs to cushion the clash of interests between management and labor. Whether it will do so in any particular case—or how fully legislation will offset the impact of innovation upon workers—becomes a political problem; but the presence of a force outside the industry, and beyond the often limited power of the labor union, introduces a factor unavailable in the late nineteenth century.

## CHAPTER FOUR

# INDUSTRIALIZATION AND OVERPRODUCTION

Overproduction was one of the most persistent themes in discussions of the American economy during the last half of the nineteenth century. Labor leaders stressed the possible dangers from overproduction, which they linked to unrestrained industrialization. The hope was to build sentiment for controls on technological innovation and thus close the gap between workers, who feared mechanization, and the mass of the public—including reformers sympathetic to the improvement of the life of working people—who defined progress in terms of technological gain.

As was true for other elements in labor's response to industrialization, the overproduction theme was common currency among trade union leaders in the 1860's, before the machine was a major concern. Mechanization only made the problem worse, but competitive capitalism was the root cause. Without the check of union rules, employers produced more goods than could be absorbed by the market, and they usually were powerless to prevent such a result, since it was competition that drove each employer to seek to outdo the others.[1] Labor leaders stressed that the results of unrestrained production were disastrous for the worker and society at large. The rush to produce eventually led to glut, since consumers were unable to buy the oversupply of goods. The final result was the periodic crisis, such as those of 1837 and 1857, which so paralyzed the economic life of the nation. Workers were idle for long periods, which not only caused privation and suffering but reduced their ability to withstand the employer's demands for lower wages. Employers found trade disrupted, and thus the injury to the worker was often matched by their own bankruptcy or losses. Eventually the entire economy ground to a halt, as in 1857, leading to adverse effects on the general population.[2]

In addition, overrapid production could also create seasonal unemployment for workers as the needs of the market were met in only part of the year. The result was to weaken the bargaining power of workers, as against employers, since wages were not high enough to build up reserves for the periods of idleness. At the resumption of production, such workers were more likely to accept whatever wages were offered in order to get back on the job.

Labor leaders stoutly denied the oft-repeated argument that an "increase in product always brings a still larger increase in demand."[3] Instead they accepted the analysis of Ira Steward, which reversed the two factors. Goods could only be sold if the consumer had the means to buy—no matter how sharply the employer lowered the cost of manufacture. In fact, if lower production costs were the result of reduced wages or the displacement of workers, the total income available to consumers dropped. Thus the lower price of the finished item was more than offset by the lower income of the purchaser.[4]

It was incorrect to define overproduction as the creation of more wealth than was needed. Instead the problem for labor leaders was twofold. First they accepted Steward's basic argument that "there never was a market so over-stocked with goods it would have taken a day to empty if destitute people had been able to pay for all they ought to have. There never was an over-production of wealth; but there has always been an under-consumption." If workers were employed fully at adequate wages, then there could be no recurrent overproduction, since "The laborer is the great employer of the laborer."[6] Second, trade union leaders also insisted that production had to be restrained as well. Stewardites rejected this argument, claiming that sufficient demand could be created, through the reduction of hours, to meet any level of production. Thus George McNeill, one of Steward's major supporters, argued in 1877 against viewing the worker as primarily a producer; in fact it was his role as a consumer that was basic to economic progress. The amount of production would ultimately be set by the demand, not by the productive capacity of industry. Overproduction was the result of society's failure to recognize that demand was not an automatic function of the cheapness of production.[7]

The objective for the trade unionists was a balance between supply and the existing demand, not unlimited production, which was supposed to stimulate sufficient demand. Labor spokesmen argued that long hours forced workers "to produce more than enough," and a reduction was necessary "to keep supply down to the requirements of society."[8] John Davis asked "why produce what cannot be consumed?" and trade unionists saw no reason to do so.[9] If demand increased in the future, under the influence of shorter hours and higher wages, production could be increased to meet the enlarged capacity to consume.

Trade unionists could offer no mechanism for establishing a balance between the supply and demand for goods; but then neither could the proponents of unlimited production, who continued to rely upon the self-regulating principles of a capitalist economy. Ira Steward broke from traditional theory in establishing shorter hours as the regulating mechanism because of its positive effect on demand. While prepared to accept this aspect of Steward's theory, trade unionists still sought the protection that a more limited production would provide for the existing skilled labor force. Trade unions tried to have the best of both worlds: they often argued for an increased consumption, but also for a limited supply. They viewed the economy primarily from their position as producers, even though they recognized that every man was also a consumer. Thus overproduction seemed more than a problem of consumption; it was also evidence of the collapse of an older labor system which the trade unions would have liked to preserve.

Labor spokesmen also pointed out that adequate consumption was imperilled by the conversion of skilled jobs into unskilled labor, often done by women and children. Such workers received significantly less in wages, and yet as machine tenders, they produced much more than adult skilled workmen. Thus overproduction could be viewed, in large part, as a situation in which the machinery worked by women and children produced too many goods for the low consumption ability of these poorly paid workers.[10] It was thus essential that child labor be eliminated and that the work of women be regulated so as to make it more difficult to substitute them for adult male workers.

By refusing to accept union regulations on apprentices and the amount of production; by pressuring workers to accept piecework; by working employees ten hours or longer when the needs of the market could be met in a shorter period, employers created a large store of goods and eventual overproduction.[11] At the same time, by denying workers fair wages, employers reduced purchasing power and insured that eventually a gap would open between the restricted demand and the soaring supply. Employers acted in this way to increase profits, but in fact they reduced them by restraining the growth of the market that could consume their goods. Labor leaders thus called for an equilibrium of production and demand, through restraint on the untrammeled growth of production and by a simultaneous expansion of demand.

In this effort, trade unions could influence both sides of the equation. Labor unions could increase demand by raising the income of workmen, but, even more directly, they could restrain the supply. As Jonathan Fincher put it: "Being unable to force a demand for our labor beyond the wants of society we should strive to keep our production within that limit, just producing a sufficiency for immediate wants, and not attempt to forestall the demand."[12] The number of apprentices could be controlled by unions as a principal means of regulating production. Some unions were able to establish their right to negotiate the output per worker, or team of workmen, with employers. However, the most far-ranging mechanism for control over the supply of goods was a reduction in the hours of labor.

Although the Stewardites envisaged the eight-hour day as the means for increasing demand, trade unionists also viewed it as a means to reduce supply. Shorter hours meant less production, and, if applied throughout the society, it would eliminate depression based upon glut by regulating supply to demand. In essence, trade unions proposed to prevent glut by less work time, rather than eliminate overproduction, once it existed, by no work time. The difference would be regularity of employment for the worker.[13]

Despite the popularity of the eight-hour day as a means to regulate production, many trade unions also sought to limit output in other ways. The most widely discussed attempt to control production in the 1860's was made by the anthracite coal miners of Pennsylvania. Overproduction had been a basic problem in the anthracite industry since the 1830's. The extensive capital needed to open and maintain a mine, especially as the search for coal led operators deeper into the earth, combined with a highly variable market because of seasonal fluctuations in demand, meant that the operator could not easily harmonize supply and demand. The large capital outlay created burdensome fixed charges that were independent of the amount produced. In addition, many operators leased their land and paid a royalty per ton of coal produced. These agreements usually specified a minimum tonnage. The operator had to produce this minimum, regardless of price, which further limited the possibility that production could be reduced. Thus the employers continued to produce, even if the prices received only covered costs or meant a loss. The result was a vicious cycle in which continued operation tended to overproduction and losses, while cessation of operations meant losses because of the heavy burden of fixed costs.[14]

The problem was not limited to American coal mines. Many of the same conditions had prevailed earlier in Great Britain. This situation was well known to American operators and miners, many of whom had come originally from Great Britain. As early as the 1830's, American operators sought to regulate production in order to stabilize the cost of coal, but until the mid-1850's these efforts were voluntary and irregular. However, in 1856, the employers in the Schuylkill region of Pennsylvania worked out a restriction plan with operators in the Wyoming fields to the north. They secured the cooperation of the coal-carrying railroads, which also owned many mines in the Wyoming area, but the plan never was tried (because of the panic of 1857).[15] The extensive demand for coal during the Civil War eliminated the overproduction problem for the moment. However, it returned with increased intensity once hostilities ended, since the productive capacity added during wartime remained while demand sagged.

The coal miners also had flirted with the idea of controlling production. Stable prices, it was hoped, would allow for steadier employment at higher wages. Restriction of production by miners was common in Great Britain, where it was a "voluntary limitation on individual production."[16] In 1848, a miners' union was organized in Schuylkill County by the English Chartist John Bates. The next year, the organization struck to reduce the production of coal, which it hoped would allow prices to rise and thus create the proper conditions for wage increases. The stoppage did reduce supply somewhat, but the union soon collapsed as a result of the opposition of employers and rumors that Bates had run off with the union's funds.[17] In 1861, the newly formed American Miners' Association suggested to the operators that they cooperate to control the price of coal. However, the employers were divided on this scheme, and nothing came of it.[18] Thus the overproduction of the postwar years was but the intensification of a longstanding problem, and after the Civil War the miners took the lead in trying to meet the issue.[19]

The end of the Civil War brought the collapse of the high wages the miners had earned during the years of conflict. The overproduction of coal seemed to be the root of the problem. In the final year of wartime demand, the capacity of the anthracite mines was 9.5 million tons. Shipments of anthracite coal increased steadily following the end of the war, and by 1870 reached 16.1 million tons. At the same time, demand for anthracite slackened because of the end of unusual wartime needs and the increased competition from bituminous coal. The sharp increase in production had been stimulated by the building of new railroad lines into the anthracite coal areas and increased purchases of coal lands by railroads to assure adequate tonnage for their lines.[20] The increased shipments, at a time of falling demand, decisively lowered prices, and the operators immediately demanded that the miners accept sharp cuts in wages. The miners resisted and, in the process, created the most important of the early coal miners' unions in the United States—the Workingmen's Benevolent Association (W. B. A.).

The initial step taken by the miners was an outgrowth of the movement for eight hours that swept the labor movement after the Civil War. Pennsylvania passed a law in 1868 defining the legal work day as eight hours but allowing additional work time should a worker and employer contract for a longer day. This proviso made the law useless, and the miners in the Schuylkill and Lehigh anthracite fields struck to gain eight hours without a reduction in pay. However, the shorter day had another objective. If miners worked only eight hours, "less coal will be shipped and what will be the consequence? Why the market will become exhausted, and your employers cannot use the old Hobby—'No Sale for Coal,' wages must come down!"[21] The shorter day thus restricted production on a continuing basis, and it served to create conditions that would raise prices as the basis for improved wages. However, the strike failed to gain the shorter day, though a wage increase was won. During the stoppage, the W. B. A. emerged as a significant power among the miners in the anthracite fields.

Beginning with 1869, the leaders of the W. B. A. made it clear that control of production was their fundamental purpose.[22] Initially this would be accomplished by strikes, although in the long term the leaders of the union called for a permanent reduction of production.[23] The strike of 1869 lasted for six weeks and did significantly reduce the supply of coal, with a resultant increase in prices. The agreement ending the strike tied the miners' wages to coal prices, thus insuring that the W. B. A. would have a permanent interest in maintaining the price of coal.[24]

Upon the end of the strike of 1869, the W. B. A. made its objective clear: "a steady healthy market which will afford to the operators and dealers fair interest on their investments, and at the same time that we may receive a fair day's wages for a fair day's work."[25] Hendrick Wright, a supporter of the miners' efforts at control of production, thought a fair price for coal would be one neither so high as to reduce consumption nor so low as to create demand that could not be met. Such a price would benefit miners and operators, but the miners could most effectively control production, through periodic suspensions of work, and thus maintain a stable price.[26] This demand for an end to overproduction certainly had roots in the history of coal mining, both in the United States and Great Britain, but it was also part of a broader effort within the American labor movement to combat the concept of unlimited production as unmitigated virtue.

The miners believed that the operators should have shared their interest in control of production. Yet many employers refused to support the proposed suspension of 1869, and the miners openly asked "Why do not the operators join hands with the association of workingmen, and introduce a system by which the market will be amply supplied but never overstocked...."[27] Although some operators in the Schuylkill region were favorable to the efforts of the W. B. A. to control production,[28] the basis for such acceptance was clearly negative. The operators' own efforts to accomplish the same objective had failed, and thus the miners represented the only alternative. However, as the W. B. A. pressed other demands, such as higher wages and the closed shop, even employers who had supported the use of the union to control production turned away from cooperation. The threat to the employers' interests from a strong union outweighed the gains that might result from cooperation on the question of output.[29]

The miners believed that the strongest opposition to cooperation came from the railroad companies that largely controlled the mines in the northern anthracite region.[30] There were two major reasons for such opposition. First, the railroad companies had to balance their interest, as producers, in a restricted supply and higher coal prices, with their interest, as carriers, in the transport of a high volume of coal. Second, the railroads were unwilling to allow any significant element in this complex balancing act to be beyond their control. Whatever the advantages in having the workers actually reduce overproduction and accept the public criticism for such restriction, the ultimate disadvantage from a strong organization of workers (which could strike over issues other than restriction, halt production, and thus reduce revenues) posed greater dangers.

In the Schuylkill region, President Franklin Gowen of the Reading Railroad reached this conclusion when it became clear that the W. B. A. would not be under his control and that it would oppose his own plans for stabilizing coal prices. Gowen planned to buy mines to secure a significant direct interest in coal production, establish pooling arrangements among the major operators—enforced by the carriers—and move towards a long-term policy of lower prices and higher volume, which inevitably would include a tight lid on the wages of miners.[31] Thus as the railroads increasingly dominated coal production throughout the region after 1870, the possibility of any cooperation between the miners and their employers disappeared.

The miners' leaders had tentative notions of labor-management cooperation that would stabilize prices, allow employers to pay higher wages, and generally reduce the conflict between operator and worker. However, these hopes were crushed by the realities of limited acceptance by even smaller employers, open hostility from the largest corporations in the area—the railroads—and divisions and lack of discipline within labor's ranks. Although the hope of cooperation with employers faded after 1871, the interest in restriction of production did not, and it was weakness of organization, rather than lack of purpose, that prevented organized efforts by the miners to restrain production after the collapse of the W. B. A. in 1875.

The depression of the 1870's added to the problems that had prompted the restriction movement of the late 1860's. Even the collapse of the W. B. A., after its defeat in the "Long Strike" of 1875, did not end the sentiment for restriction of production. A convention of miners met at McKeesport, Pennsylvania, in early 1876, and resolved in favor of the eight-hour day and a reduction of production per man per day.[32] In 1877, John M. Davis, Editor of the *National Labor Tribune,* pointed out that the major employers had agreed to a pooling arrangement that set maximum production at 8 million tons of anthracite. He urged the miners to use the same principle of restriction to make sure that overproduction would not lower prices and wages. So vital an issue for workingmen should be under their control. However, the requisite organization was lacking, and thus little could be accomplished.[33]

Among the employers, efforts continued in the 1880's to control the overproduction of bituminous and anthracite coal. However, in the anthracite fields, firm control over output was not effected until 1900, when the Reading Railroad secured control of 63 percent of the deposits by buying out the Central Railroad of New Jersey and negotiating perpetual contracts governing the production by independent operators.[34] The miners also continued to call for restriction as one of their basic aims. The constitutions of the miners' organizations of the 1880's included explicit calls for restriction.[35]

At the convention of bituminous miners in December 1889—which was preliminary to the organization of the United Mine Workers the next year—President John McBride stressed the importance of regulating output to prevent overproduction. Three methods were available. The first would assign "a certain tonnage per day to each miner" in order to limit supply to the demands of the market. By such a policy the miners "would speedily become masters of the situation and be able to command better pay. . . ." Second, the hours must be reduced to eight per day. McBride was realistic enough to note that both these methods had met with considerable opposition from within the ranks of the miners because they meant an immediate reduction in wages (most miners worked at piece rates) in expectation of future gains. Men living at the poverty level found it especially difficult to make this exchange.

The third method available to the miner was the one used in the 1860's. McBride noted that general suspensions to reduce output affected other industries and led to public hostility. Yet if the suspension was an effective device in controlling overproduction, McBride insisted it would have to be carried through despite public opposition. The convention adopted a resolution "in favor of a restriction in the output of coal," but left further action to the new United Mine Workers.[36]

Despite this resolution, little effective action for restriction developed in the 1890's. The large number of nonunion mines insured that any plan for restricting production of bituminous coal would excite the most severe opposition from employers in unionized mines. They would be forced to limit production, while their nonunion competitors produced to full capacity and thus gained maximum advantage from any rise in prices that did occur because of the suspension. Also the problem of controlling production was further complicated, after 1890, by the expanding use of coal-cutting machinery that increased productivity per miner by division of labor.[37] By the end of the century, little had been done to translate the sentiment for restriction into effective action. The continuing factors of an imperfectly organized industry, the opposition of employers, and the reluctance of many miners to give up any portion of their scanty present wages for the supposed future gains from restraint of production offered little hope to the leaders of the United Mine Workers that control of overproduction could become a reality.

Underlying labor's position on overproduction was a refusal to embrace unlimited output as the source of all social progress. Production had to be limited, in the first instance, by demand; but more basically production was the servant of the populace and not its master. Albert M. Winn, President of the Mechanics' State Council of California, argued the point boldly in opposing Chinese immigration in the late 1860's. To the argument that such workers were needed to develop the resources of the nation, he answered that "There is no good reason for developing faster than the happiness of the people required—anything more is a fallacy." In this case, the interests of the existing labor force would be undercut by the cheaper workers from China, and thus Winn was clear on the need to restrict development, if such were the actual result.[38]

Labor leaders totally rejected the concept of progress as an abstraction: it had to be defined in terms of the effect upon existing groups in society. Innovation took place because of the concrete benefits perceived, not for any intangible reasons. The pace of economic change in the American economy was fueled by the profits expected by entrepreneurs. It was not that workers opposed progress, and employers favored it, but that the type of progress carried through subordinated the interests of labor to those of capital. Trade unions represented a different set of interests, which could be best served by substituting the concept of a balanced, fair economy—one that considered the established rights of workers—for the efficient economy that was basic to industrialization. The overproduction thesis seemed proof to labor leaders that their set of economic relationships was not only best for workers, but for society at large.

Labor leaders claimed that ultimately profits depended upon the earnings of the work force. Thus progress had to be reconciled with the interests of those already at work. Unfortunately, the competitive capitalism of the period and the absence of state action as a viable alternative made it difficult to advance with maximum protection for the interests of the established labor force. Lacking a means of adjustment, many trade unions sought restraint with little regard for its effects upon the industry, and many employers demanded innovation with little regard for the effects upon their workers. The employers generally won this test of strength. Had the trade unions been able to carry out their objectives, the nation would not have been so thoroughly and quickly industrialized, nor so rapidly populated, nor so completely developed, nor so wealthy in terms of gross national product; whether it would have been less happy for such a history is not so clear.

The long depression of the 1870's focused increased attention on overproduction, which, in turn, was clearly related to the increased use of machinery. Mechanization threw men out of work and thus cut into consumption. At the same time, it increased the supply. The gap widened until goods glutted the market, leading to the collapse of the economy. The solution was to reduce hours in proportion to the increase in productivity of machinery, so that the labor force would be fully employed and purchasing power would remain high.[39] Although eight hours was the standard generally proposed, it was not to be considered a final figure. Eight hours was the first step. It was essential in order to catch up with the overproduction already generated by improvements in the methods of production. As technology improved, further reductions in hours would be needed to continue the balance between supply and demand.[40]

It was also important to end the established system of adjusting production and consumption, which varied the number of workers rather than the number of hours.[41] Trade unions feared an increase in the supply of labor almost as much as a reduction, since there was no guarantee that all those employed in a prosperous year would not be turned into competitors for work in a slower period. However, a flexible work day would lessen this fear about an oversupply of workers.

The coopers were one of the trades most affected by mechanization before 1873, and thus it is not surprising that a vice-president of the Coopers' International Union, Robert Schilling, should focus on the responsibility of the machine for overproduction. He argued that the periodic economic crises in the United States were the result of overproduction stimulated by mechanization. Schilling acknowledged that machinery cheapened prices and thus enlarged the market, but he wondered whether such an increase was large enough to use up the vastly greater supply. Moreover, lower prices hurt many businessmen and led to failures that further affected the economy. Schilling denied that he was opposed to machinery, but he called for a reduction of hours to offset the negative effects of machinery upon workers, and, in general, to strengthen the economy.[42] This line of thought was central to discussions of overproduction in the 1870's. The most concerted effort to control production in the 1870's occurred among the lamp chimney workers of Pittsburgh, who struck for over two years in an effort to prevent the loss of traditional restraints upon output.

The glass chimney was an important component of the oil-burning lamp in the late nineteenth century. Until 1876, highly skilled glass workers made lamp chimneys by hand. The process first required a "gatherer," who collected the proper amount of molten glass from the furnace on the end of a long hollow tube and blew it into the shape of an elongated pear. A "blower" then blew the glass into the final desired shape and size. The hollow glass chimney still had to be opened at both ends, and this work was also done by the blower, using hand tools. The heel of the chimney was opened, smoothed, and shaped to fit the lamp body itself, and the top was opened and polished.[43] The work required a high degree of skill, which allowed the workers, as early as the 1860's, to set specified numbers of chimneys for each "move" or half-day of work.[44] In 1875, a crimp top chimney was introduced. The skilled glass worker crimped the top of the chimney, instead of simply polishing it. Since more time was required for the additional operation, the number of chimneys to be made in each move was reduced.[45]

The advantages of a specified output were enormous for the worker. Not only was there protection against overproduction, with the resulting unemployment and downward pressure on wages, but there was also a uniformity of labor costs for all unionized shops. In the highly competitive lamp chimney industry, such uniformity reduced the possibility that employers would use lower costs as a means for gaining an advantage over competitors. With output, and the corresponding wages, standard, employers had to look to other aspects of their business in seeking to lower costs. The introduction of machinery was the result.

In 1876, a Pittsburgh firm introduced a simple machine that crimped the top of the lamp chimney. The "crimper" was merely a circular crimped mold with a revolving cone inside. After the top of the chimney had been opened, it was softened by heating and placed in the crimper. The revolving cone smoothed the top of the chimney and guided it toward the mold, which crimped the glass.[46] The machine was operated by an unskilled worker, often a boy. This simple device triggered a series of events that culminated in a strike of the skilled lamp chimney workers.

The danger of the crimper to the skilled workers had little to do with displacement from the trade. The lamp chimney still had to be blown in the traditional way. The crimper merely quickened the final step of the process. Instead the problem was that the crimper allowed more lamp chimneys to be made, per move, by taking the final crimping from the skilled worker and giving it to an unskilled worker operating the machine. The Fort Pitt Glass Company, which originally had exclusive control of the crimper, demanded an increase in the allotted number of lamp chimneys to be made in a move, without any increase in pay for the skilled workers. The machine was thus a device for reducing the labor cost per lamp chimney—a most attractive possibility in the midst of the slack demand occasioned by the depression of the 1870's.

The workers at the Fort Pitt Company accepted the increase in production per move. However, the local union of lamp chimney workers rejected the spread of the crimping machine to other shops.[47] The Fort Pitt Company soon lost its exclusive rights to the crimper. As the device became available to other glass firms, the employers demanded an increase in output, without an increase in pay. The glass workers refused this demand, and the strike began in June 1877.

The union's opposition was based upon two considerations. First, the market for lamp chimneys was not an expanding one in the 1870's, and thus an increase in output per day would most likely lead to overproduction and some unemployment among the skilled workers.[48] Second, and more important, the crimper threatened to wreck the stability of cost, which had been the basic protection for the workers against downward pressure on wages. The Fort Pitt Company had insisted on an increase in output, based on its use of the crimper, with no concern for the principle of standard output. Other employers had responded by seeking to increase output by hand methods. After the strike began in June 1877, the very principle of standard output collapsed as employers sought to use labor cost as a competitive tool.[49] In defense of the old system, the workers pointed out that continuance of a specified output meant that "The manufacturers will know what to expect each turn, and what each article will cost."[50] Thus the union insisted that the manufacturers and the workers continue to restrain competition through negotiated limits on output.

In sum, the basic issue was whether an increase in production—in this case as the result of a mechanical innovation—could be put into effect without the concurrence of the workers. The skilled glass workers could have lived with the crimper, but they opposed both unlimited output and any increase in production without a proportionate rise in wages. If employers had to share the benefits of higher productivity with the workers, they would be less prone to increase output, thus reducing the possibility of overproduction.

The strike dragged on for two years, which was unexpected considering the fact that glass centers beyond the Pittsburgh-Eastern Ohio region were not affected. It was not until February 1879 that the employers claimed they had been able to replace the strikers and resume operations.[51] Ultimately, the return of prosperous times made both workers and manufacturers anxious to end the strike, and so in July 1879 an agreement was concluded. The crimping machine was to be used wherever the employer desired. However, another newly invented machine, "the opener," was barred. This machine was designed to speed the finishing operation on the heel of the chimney. The employers conceded this point because the opener was still the exclusive property of two firms, and thus the majority of manufacturers were as eager as the workers to prevent its use at that time. The employers tried to protect themselves from the Eastern glass manufacturers by a proviso that the opener might be introduced "if its use is necessary to a successful competition with the Eastern trade." The workers further had to accept the higher output per move originally demanded by the manufacturers without an increase in wages.[52]

Despite these obvious reverses, the workers did not suffer a crushing defeat, since the principle of regulated output was retained. Clearly many employers were also opposed to continuing the unrestrained competition and chaotic business conditions that marked the period of the strike. They demanded the right to introduce the crimper and thus increase productivity but were as unwilling as the workers to abandon both uniformity of production and limits on total output. If costs of production could be lowered by mechanization, but output remained uniform and limited to the levels that insured the most advantageous prices, profits could be maximized. Thus the workers had to produce more for the same wages and admit the use of the crimper, but they salvaged the principle of a standard output per day and barred the opener, which might have exacerbated the situation still more. Although employers modified the situation in their favor, the fundamental structure of the trade remained the same.

Restrictions on output continued to be popular with skilled glass workers until the 1890's. The window glass workers exerted the most effective control over production because "The supply must be kept equal to the demand. It must never exceed it."[53] These restrictions meant less in wages for the workers, who were paid by the box, but it prevented the surplus production and competition among employers that could mean far greater losses in wages over the long term.

In the manufacture of prescription bottles, frequent strikes were the rule during the decade following 1873. Employers claimed that the trade was dull, and thus they demanded reductions in wages. However, the workers decided to strike rather than accept the reduction. "It would be better to lie idle two or three months, and allow the surplus products to be wiped out, than to continue to work at the reduced rate of wages and have the surplus on the market. In other words, the reduction of wages would not be beneficial either to the men or the manufacturer, but the idleness was; it allowed the surplus product to be wiped out by the power of consumption."[54]

The Flint Glass Workers had restrictions on production in the 1880's, and the Green Glass Workers tried to establish such limits in the depression year of 1894. However, the policy of restriction lost appeal within the glass workers' unions in the 1890's because of its weakness in the face of nonunion plants that ignored such limits. The competition from unorganized factories placed unionized companies at a competitive disadvantage, leading them to demand that the union's restrictions be modified.[55] In the window glass trade, the refusal of Local Assembly 300 of the Knights of Labor to accept modifications of its rigid limits on output led to dual unionism. The competing Window Glass Workers' Association was prepared to supply workers who did not accept the production limits negotiated by the employers' association and Local Assembly 300.[56] In addition, skilled union glass workers became jealous of those who worked in nonunion plants. Although nonunion workers were paid lower piece rates, they earned more because of higher production and longer periods of employment.[57]

Thus the trade unions in the glass industry discovered that their policies foundered on an inability to organize the entire trade. Failure to do so, combined with the nationalizing of markets so that nonunion plants in one area of the country became competitors of union concerns in another area, put pressure on the unions to modify their policies of restriction. Control of output was designed to avoid overproduction, curb competition, and thus make it less likely that lower wages or the layoff would become tools of the employer in his search for a greater share of the market. Yet such a policy was viable only so long as there were not firms that could ignore the restrictions and undercut unionized companies. In fact, stability in the trade, based upon restriction of output, became an invitation for entrepreneurs to enter the industry, operate nonunion facilities without limitations, and thus gain a competitive advantage. Unless the union had substantial control of the work force, new companies could not be stopped from exerting pressure on the existing arrangement in the trade. Since most unions lacked such control, restrictions of output had only limited success as a weapon against the insecurity and destructive impact of industrial change.

While trade unions often focused their attention on controls over output, discussion continued, both inside and outside the labor movement, concerning the role of insufficient demand as the primary reason for overproduction. Charles Litchman, of the Knights of Labor, asked rhetorically, "If the naked were clothed, the homeless sheltered, the hungry fed, would we hear of this thing termed *over-production?*"[58] As John McGrath, the Commissioner of the Michigan Bureau of Labor Statistics, put it: "Our productive capacity has been unlimited, our consumptive capacity has been limited."[59]

Yet the acceptance of such a conclusion could lead to little more than useless tinkering that would make no significant change in the situation. Thus the United States Commissioner of Labor, Carroll Wright, embraced the underconsumption argument heartily in 1898, but felt that industrial education would raise the wages of the poorer laborers to the levels of the better paid.[60] Such a conclusion ignored the fact that education and skill were hardly a viable base for increasing consumption when mechanization destroyed the need for special training as the basis of production. Wright's blindness on this point undoubtedly flowed from his inability to accept those changes that might truly increase the consuming power of the people.

One such change was support for trade unions. The reform economist, Richard Ely, argued that trade unionism was important to the health of society precisely because it protected the legitimate interests of workers. James Means, a leading New England manufacturer, spoke for that small group of employers who recognized that their own profits, and the continuance of American capitalism, depended upon some restraint on the tendency of business to destroy its long-range interests in pursuit of its short-run goals. Thus Means accepted the underconsumption argument as the reason for the glut of goods that threatened the prosperity of the nation. He argued that this lack of purchasing power resulted from the inequality between worker and employer in wage bargaining, and the necessity for employers, in a competitive situation, to reduce wages to the minimum paid elsewhere. Labor was not paid a fair wage, but a market wage, and trade unions were the instrument by which the bargaining power of workers and employers was made more equal. "Orderly trades unions" should be encouraged. The higher level of wages that they were bound to gain would increase purchasing power and eliminate the overproduction that so upset the profitable operations of business. Also Means pointed to the uniformity of wages that was a primary objective of unions. Success by a trade union in this area would eliminate one of the major sources of cutthroat competition. Thus unions served a positive function, within a mature capitalist system, by insuring that the consuming power of the worker would not lag behind the productive gains of industrialization.[61]

Unfortunately for the trade unions, most employers rejected Means's analysis in favor of the traditional one that stressed cheapness in the cost of production as the lever of any mass consumption economy. Mechanization produced a great bulk of goods at less cost and thus allowed workers to buy items once available only to the rich.[62] A shorter work day was possible only to the extent that it did not reduce productivity per man-hour and thus increase costs. Higher wages were likely to raise costs as well, and so were a negative factor unless accompanied by proportionate increases in productivity. In such an argument, workers basically improved their standard of living through gains as consumers. Basing economic progress on unlimited and cheap production meant that the interests of skilled workers in the preservation of their jobs and conditions of labor had to be sacrificed, not only for the profit of the employer, but the benefit of the mass of the people.

In attempting to protect workers against the effects of industrialization, labor leaders faced the general support for mechanical and technical change among the American people. The labor movement had hoped to find some means of convincing the public, and even employers, that uncontrolled industrialization hurt their interests as well as those of the existing skilled work force. The persistent stress on the overproduction thesis was a major mechanism for accomplishing this. However, it proved to be a frail weapon against the worship of machinery, even though it won considerable acceptance among reformers, moderates, and even some employers and conservatives. One could accept overproduction, but not organized labor's conclusions about its origins, nor the solutions offered by trade unions to offset its effects. Thus the overproduction thesis did not become the basis for a viable and widely accepted alternative to unrestrained industrialization.

## CHAPTER FIVE

# EARLY RESPONSES TO TECHNOLOGICAL
# CHANGE BY TRADE UNIONS

In order to understand the response of the American labor movement to industrialization, it is important to recognize that the reaction took place on two distinct levels. First, organized labor supported general measures designed to limit the impact of industrial change. As we have observed, the overproduction issue was heavily stressed in an effort to slow the rate of innovation, and also as a basis for demands that the workday be shortened. The labor movement also focused attention on the host of social ills, including crime and pauperism, that flowed from unemployment, and mechanization was identified as the major cause of idleness. It was a clear effort to link the interests of those displaced with the concerns of the general community. If this were done successfully, the possibility increased that remedies such as the abolition of child labor, the restriction of female labor, and the limitation of immigration—which supposedly would reduce the displacing effect of mechanization—could win public support, especially when other groups wanted the same reforms for other reasons. Thus the labor movement sought to avoid isolation from the community at large by ultimately accepting innovation and seeking measures that would cushion the ill effects upon the worker.

Second, trade unions had to face the direct impact of innovation within their own industries. Under pressure from technological change, workers initially were likely to share the sentiments of the reformer William West, who described "labor saving machines as curses, inventors as plagues, and patent laws as embodiments of tyranny."[1] The skilled worker, throughout the nineteenth century, believed that his skill was the key to his economic position as well as his status, and unlike disputes over hours and wages, mechanization threatened that basic skill and thus the very foundation of his position in society. Hostility to machinery could be expected, and hostility there was, but usually with a solid understanding of the limitations that American society imposed. Thus the initial calls for opposition to technological change were soon modified by the conditions of the trade, the limits imposed by private property, and the reality of a largely unregulated American system of capitalism, which viewed innovation as a basic decision of employers. In addition, the popularity of progress as a generalized ideal and the specific attraction of mass production to those citizens who could profit as consumers further limited the urge to fight against technological change.

The reaction of trade unions to innovation became a vastly complicated task of balancing the leverage that skill and the organization of workers would give to the artisan and his labor organization against the powerful factors that favored industrial change. Different choices of action marked specific trades, and success or failure was dependent on how well unions and workers read the relative influence of their own skill and organization, the economic advantages to the employer of the innovation, the rate at which changes would be introduced, and the usefulness of skilled workers in the new work situation as against the feasibility of using women, children, or unskilled men. Most unions recognized that it was futile to oppose innovation in their industry, and instead they sought control of the new jobs created by industrial change as well as other measures designed to protect the skilled worker to the extent possible. This was the course of action taken by the cigar makers and coopers as early as 1870.

The Cigar Makers' International Union went through three distinct periods in its response to technological change, and an examination of the changes in policy reveals the complex factors that determine the course of reaction to innovation within a specific trade. From the introduction of a mechanical device in 1869, until 1873, the union sought a single policy for all members and decided upon acceptance of innovation with certain significant restraints; during the depression of the 1870's, it acquiesced to innovation; and from the late 1870's until the 1920's, the union maintained a system of local option that allowed the particular needs of local units to determine policy, with the result that a significant portion of the cigar makers opposed innovation.

Until the introduction of the mould in the late 1860's, cigars had been manufactured in America by skilled workmen who made the product from start to finish. The cigar maker first formed bunches of tobacco that comprised the filler of the cigar by arranging and shaping the individual pieces. He next placed this filler into a binder leaf that held the body of the cigar in shape, and subsequently put on wrapper tobacco to produce the finished item. The mould was nothing more than a piece of wood with "moulds" shaped in it to accommodate the amount of filler tobacco needed for a specific type of cigar. The moulds were then placed in a hand press that formed the tobacco under pressure into a proper bunch that could then be wrapped to complete the cigar.[2] The mould was thus a tool to supplement hand labor, not a machine; but the danger to the skilled cigar maker lay in the division of tasks that accompanied its use.

Where moulds were introduced, employers generally hired "filler breakers," whose job it was to make the filler for the cigar through the use of the mould process. Since the process of preparing the filler was simplified, unskilled labor, including women and children, could be used for this task, and one employer estimated that he could train filler breakers in two weeks rather than the three years required for apprenticeship by the Cigar Makers' Union. These less-skilled filler breakers were paid less per thousand cigars, though employers claimed they could earn as much as cigar makers, working without the mould, because of the increase in production.[3] Thus productivity increased, and piece rates fell, a situation that unionists have always distrusted because of the possibilities for a speedup. The wrapping of the filler and the final finishing of the cigar were still done by skilled workers (now called rollers) although the element of skill was somewhat reduced, since the filler from the mould was easier to wrap than the one made by the older process of hand shaping. Thus the mould divided the production process, lessened the skill involved throughout (but so greatly in filler breaking as to allow unskilled workers to enter the trade), and increased productivity. These results recommended the new device to employers. The immediate threat of the mould process forced the cigar makers to decide how to meet the problems posed by this innovation.[4]

One alternative, of course, was to resist the introduction of the mould. Locals of the Cigar Makers' International Union struck against the use of the mould, and support for this position came from officers of the national union. A Toledo cigar maker urged opposition to the mould even though European workers had tried to oppose its introduction without success. He regarded European workers as only "a grade above the slave" whereas American workmen would assert their rights. "The question of Moulds must be met with *now*. We must form in the line of battle, our face to the front, our foe, the Mould, and our battle cry, *unconditional surrender,* for this will result in the death of our trade or the death of the Mould system. . . ."[5] At the union convention in September 1871, President Frederick Blend argued that the mould allowed division of labor, which threatened the skill of the cigar maker and the bargaining power of the union. He claimed that strikes against the mould had been successful, and he urged total opposition.[6] However, Blend's was the minority position within the union.

The majority accepted Blend's fear about the division of labor but rejected the feasibility of total opposition to the mould. At the convention of 1871, the committee on the president's report specifically disagreed with Blend's attitude toward the mould, since if the moulds were a success they would be worked "our objections notwithstanding."[7] As a New York cigar maker had pointed out earlier in the year, refusal to accept a successful innovation only meant that nonunion men would be employed. Instead, the organized cigar makers should gain control of the new innovation, which would make it subject to union regulations.[8] The union thus adopted a policy of accepting the mould, but not the division of labor.[9]

The threat to the cigar maker was not from the mould *per se,* but from the use of the mould to divide the process of manufacture so that unskilled labor could be substituted for the skilled worker. The union thus adopted a policy that permitted the mould so long as it was used without the division of labor into filler breakers and rollers. One skilled cigar maker would continue to manufacture the entire product, but at somewhat greater speed because of the use of the mould in place of hand shaping of the filler. Selig Perlman viewed this as a policy of opposition to the mould, since the mould was most efficiently used by dividing the work into filler breaking and rolling.[10] Actually, it was a middle position between those who supported obstruction—and were determined to block the use of the mould under any conditions—and those who would have accepted the division of labor, with the subsequent admission of filler breakers into the union. President Edwin Johnson argued for this latter position at the convention of 1872. He viewed the use of filler breakers as the most effective way to utilize the mould, and thus employers insisted on this procedure. Since unionized cigar makers were prohibited from accepting employment in a division of labor system, they were barred from jobs in many shops. As a result, members were leaving the union. Johnson was prepared to admit the evil of division of labor, with the resulting dilution of skill, but he claimed that since most cigar makers were not members of the union, and most were ready to work in a division of labor situation, it was foolhardy for the union to oppose. Instead it should accept the division of labor and organize all cigar makers.[11]

Johnson's position was rejected in the same way as Blend's had been a year earlier. Instead the union's policy was to accept the mould, but not the division of labor. The national union made this distinction in its attitude toward strikes by locals. Strikes against working with the mould, or strikes to make the use of the mould uneconomic by demands for higher prices on goods made by the new process, were not authorized by the national executive board. The result was to bar aid from the national union in the form of strike benefits. However, strikes against working with filler breakers received financial support from the national union.[12]

An attempt to reverse the policy in the direction of greater opposition—through permitting full local option on the matter of moulds—had been defeated in 1871 by a vote of 33 to 5.[13] President Johnson's attempt to interpret the constitution of the union so as to permit filler breaking, so long as union men were the filler breakers, was rejected by the convention of 1872 by a vote of 18 to 2.[14] Thus a majority of the union delegates strove to maintain the policy of opposing the division of the trade, regardless of who was employed, but of accepting the mould without a division of labor.

Future experience would indicate that few employers would use the mould without a division of labor, since the productivity gains under the union's system were negligible when compared with the use of filler breakers and rollers. By 1880, the Cigar Makers' Union accepted this fact; but in the early 1870's the future of the mould was still in doubt, and thus a majority of the union leaders based their policy on what would prove to be a vain hope in the compatibility of the mould and the existing system of skilled labor.

This policy was pushed to its practical limits at the convention of 1873 when the constitution was amended to permit union cigar makers to work in shops where filler breakers were employed, so long as they did not "work in conjunction with a filler breaker."[15] The objective was to expand employment opportunities for union cigar makers in those shops which might produce some types of cigars by a division of labor but still were prepared to accept the union's mode of working for other cigars.

The key continued to be a refusal to accept the mould when worked with a division of labor. This could be called opposition only if one defines that term as innovation that serves the employer's interests without reference to the needs of the workers. Filler breaking meant peak efficiency from the use of the mould and maximum profits for the manufacturer; but it also constituted a real threat to the standards of labor that had been built up under the protection of skill and the resulting hand process of manufacturing cigars. The division of labor meant that unskilled workers, particularly women and children, would become the filler breakers, forcing skilled men into the rolling specialty. Even though one filler breaker could supply two rollers, the concentration of skilled men in only one specialty of a divided trade would tend to reduce the bargaining power of workers as against the employer. Thus displacement as such was less the issue than the fear that a division of labor would reduce the bargaining power of the skilled men still left in the trade.

In other trades, employers sought to use division of labor for the same purpose, often combining a division of labor with a demand for more apprentices, who might be easily trained for the simpler jobs now available. This would increase the supply of available labor. The reaction of the cigar makers in refusing to accept the division of labor can be equated with the opposition to the loosening of apprenticeship rules in other trades, even though it was quite apparent that apprenticeship as such was a dying institution. Employers sought to increase the supply of workers, and the division of labor was essential to this purpose. Workers opposed such plans, for they regarded a liberal supply of labor as giving a bargaining advantage to the employer.

The mould, used together with filler breakers, would have greatly reduced the ability of skilled cigar makers to bargain effectively by virtue of the scarcity of their skill and the ability of the union to organize a significant number of these skilled workers. The Cigar Makers' International Union insisted that the protection offered by skill and organization not be lost. The union accepted innovation, but not unrestrained innovation; it was prepared to acknowledge the advantages that the mould would offer in increasing productivity—with the consequent advantages to the consumer and the increased profit for the employer—but not without protection of the worker's vital interests.

In sum, the cigar makers accepted technological progress that respected their needs as workers, but they opposed innovation that served only the employer or the consumer. Most trade unions supported such a policy, but the private enterprise system of the period stressed a conflict between the interests of labor and the employer, and there was no available instrument for a policy of regulated innovation. Thus employers used technological advances to reduce the bargaining power of their workers through the elimination or weakening of skill. In such an atmosphere, the Cigar Makers' policy of restraining the effects of innovation in order to protect the existing conditions of labor more fully cannot be simply labeled as opposition. It refused to deify unrestrained productive efficiency and insisted on consideration for those who were in the existing labor force. In a different age, one could see the possibilities for compromise and for restrained technological advance; but in the late nineteenth century such accommodations were rarely made in America.

The unemployment that followed in the wake of the panic of 1873 led the handful of delegates who appeared for the Cigar Makers' convention of 1875 to change the existing policy. The convention permitted filler breakers to join the union and thus acknowledged the division of labor. John Junio of Syracuse, a former president of the International Union, proposed the change, and he echoed President Edwin Johnson's support of this position in 1872. Junio insisted that only unionization of the entire labor force in the cigar making industry could "control the surplus labor which is proving ruinous to the skilled workman."[16] In the depressed trade, the opposition to division of labor meant that the union men were barred from precious jobs because filler breaking was used in conjunction with the mould. Whatever the desirability of the defense of an undivided craft, it had to yield to the practical necessity that even work in a shop with division of labor was better than no job at all. Also, the admission of all cigar makers was necessary to save the union itself. In 1875, President William Cannon despaired of the union's continued existence. Strikes against wage reductions had been lost, and the result was often the collapse of the local.[17] The prohibition against working with filler breakers led other union members to leave to take whatever employment they could. Finally, the new provision would allow recruitment among an increasing body of workers who had been barred from membership. Thus practical necessity set the limits on the policy of restraining division of labor in the trade.

With the end of the hard times of the 1870's, the Cigar Makers' International Union revived, and it quickly developed a new policy on innovation that lasted until the 1920's. By 1881, membership had tripled from the low point of 1877, and skilled cigar makers once again could expect reasonably steady employment. Thus the provision accepting filler breakers into the Union was dropped, and the national union gave up trying to adopt a single policy on innovation for all of its members. Instead the admission of filler breakers, rollers, and other workers in shops that used a division of labor was made "optional with the local union, except in places where the system has already been introduced."[18] Since locals determined whether a division of labor had "already been introduced," and since it was clear by the 1880's that the policy of the early 1870's could not succeed, because most employers would not use the mould without a division of labor, the local option policy actually permitted locals to oppose any innovation without forcing such a policy on other locals. As a result of local option, division of labor was permitted in New York City, Philadelphia, Baltimore, and Cincinnati, but barred in Boston, Buffalo, Cleveland, Tampa, Chicago, and San Francisco.[19]

This decision has been cited as an example of a suicidal policy of obstruction to innovation.[20] It certainly did lead to a decline in the membership of the union, but it was reasonable when viewed from the perspective of the cigar maker who worked in the portion of the industry that continued to make relatively expensive handmade cigars. These unionists could continue to restrain division of labor because their employers were willing to accept hand production, in large measure because such methods meant a finer product, which would be a selling point for the high-priced cigar. The mould increased productivity by one-third, since it allowed the filler breaker to form the body of the cigar with much less attention to the amount and position of the tobacco. However, the result was that the filler could contain too much tobacco, and the mould would then produce hard spots that did not smoke well. Also the filler from the mould was dried before the wrapper was rolled on, so that it did not cling as tightly as in the product made in one operation.[21] Thus the study of cigar making by the United States Commissioner of Labor in 1904 concluded that "no other cigar is as good as the one carefully made by hand. . . ."[22] The handmade product might be of higher quality in other industries as well, but the handmade cigar was still able to command a market, since the costs by hand were not prohibitive and the advantages in hand production were widely recognized.

Accordingly, opposition to division of labor was feasible for those cigar makers who anticipated employment in that segment of the trade that continued to produce finer cigars. However, the determination of whether to continue opposition to division of labor rested with the local, allowing those locals that were engaged in the manufacture of cheaper cigars—where division of labor was the predominant method—to accept innovation and unionize filler breakers, rollers, and other cigar workers. Most craftsmen hoped that hand production would continue even after innovation had divided labor, but cigar making was one of the trades in which such actually was the case. This allowed the skilled cigar maker the option that other artisans did not have once machine methods were viewed as more profitable.

A policy of opposition to division of labor could not work where the use of the mould with filler breakers, and later the cigar making machine, were deemed necessary by manufacturers to a profitable business. Thus the competition of cigarettes, in the early twentieth century, led to the development of cigar making machines that were indispensable in the production of the five-cent cigar that could compete with the newer form of tobacco. It was the preference for cigarettes that cut into the ranks of cigar makers in the period 1900–1930 as much as the policies of the union toward innovation.[23] As the production of fine cigars declined, and as it became a smaller portion of the trade, the skilled hand cigar maker faced the choice of accepting the mould system and the cigar making machine, which opened the union to semiskilled and unskilled workers, or maintaining his skill by keeping out the division of labor and the machine. This latter policy meant that the union would control only the production of cigars that could still be sold profitably, even with hand methods of production, and thus the union's membership would decline steadily as a proportion of the total work force in the industry. Despite the urgings of the national officers, the skilled cigar makers, who controlled the union, chose to preserve the pockets of hand work by retaining the right of a local to refuse to organize workers who operated machines or made cigars by the mould process.[24]

This was less a policy of suicide than a calculated attempt at preservation by a minority of skilled workers in what had become a basically mass production industry. In other trades, hand methods were not able to compete with technological changes, and thus the options available to skilled craftsmen were greatly reduced. The experience of the coopers illustrates this situation well.

The manufacture of barrels was a handicraft until the 1860's, when the discovery of oil in Pennsylvania greatly expanded the demand for the wooden barrel. Employers sought to mechanize the manufacturing process almost immediately, in large measure to weaken the bargaining position of the cooper, who was regarded as an undependable worker. Employers complained about the cooper's hard drinking, and the resulting "blue Monday," as well as his independence and itinerant nature, which made it difficult to maintain a steady supply of skilled workers.[25] As early as 1860, machinery was on the market to cut the staves and heads of the barrel, thus allowing the cooper to assemble these precut components rather than build the entire item.[26] By the end of the decade, these components were widely made by machinery. However, the demand for oil continued to grow, and the wooden barrel still served as the means of transport. In the early 1870's, the Standard Oil Company thus took the lead in developing machinery that would quicken the assembly of the barrel by reducing the amount of skilled hand labor.[27] The use of this new machinery drastically affected the cooper, for it allowed the basic assembly of the barrel to be done by unskilled labor, with only the placement of the heads and the hoops still done by hand.[28]

The results of these technological innovations were dramatic. The machinery reduced the need for skilled labor sharply. One cooperative cooperage shop in Minneapolis, which manufactured barrels for flour without the assembly machinery from 1874 to 1885, was finally forced to mechanize to meet competition, since machine-made barrels were cheaper. The result was a reduction of coopers from 120 to 90 and the addition of 20 unskilled men and boys.[29] In noncooperative shops, the displacement rate was undoubtedly greater as employers sought to reduce the need for skilled labor to a minimum. The long hours in the trade, which exceeded those among most skilled workers, also cut into opportunities for employment.

In addition, the process of mechanization created a division of labor, which reduced wages for the formerly skilled worker, even were he to find employment in the machine shop. A cooper who worked in a machine shop in the early 1880's reported that a surplus of hands, piecework, and breakdowns of machinery prevented average daily wages from exceeding $1.50.[30]

This was far below the $3.00 per day that skilled hand coopers had received in 1880. In 1898, a study by the United States Commissioner of Labor revealed that wages for the semiskilled work in machine barrel shops in 1895 were generally between $2.00 and $2.50 per day, once again below the level expected by skilled workers in 1880. Unskilled machine tenders in these shops did not receive more than $2.00 per day in 1895.[31] Unlike the cigar trade, no sizable portion of the product was still made by hand after the 1870's. By the end of the century, the last hand processes in the manufacture of barrels had been mechanized, and the last functions of the skilled cooper were gone.[32]

The coopers were strongly opposed to the introduction of machinery, even in the 1860's, when the assembly of the barrel itself was still a hand process.[33] This hostility to the machine led to the formation of the Coopers' International Union in 1870.[34] The coopers did not deny their opposition to the machine. In clear terms, they opposed machinery because it served only the interests of the employers at the expense of the craftsmen. There were many unemployed coopers available who could meet the enlarged need for barrels, and thus an increase in production did not require machinery. Employers were attracted to machinery because it allowed them "to make money and enslave coopers" by superseding the skilled worker.

A successful strike against the introduction of machines in Allegheny County, Pennsylvania, was applauded, and others were predicted.[35] Labor-saving machinery should lighten the toil of workmen, not displace them in favor of unskilled men or boys.[36] Such displacement was of no value to society at large, since it increased unemployment and low wages—the very conditions that produced dishonest, intemperate, and slavish men.[37] Machinery also was inimical to the long-run interests of the employer, since it greatly increased production and led to an excess of supply that lowered prices. Production had to be restrained so that demand for barrels would equal or exceed the supply, leading to higher prices for the employer and ultimately higher wages for the cooper.[38]

Despite the strikes against machinery, the leaders of the Coopers' International Union soon recognized that opposition could not block the introduction of machinery. Thus the emphasis switched from blocking machinery to meeting its effects. Shorter hours became the major vehicle for combating the destructive results of mechanization. The coopers argued that a shorter work day would mean less output per worker, and this would help to create an equilibrium between supply and demand that would raise prices. It would also increase the number of coopers who would be employed. These two developments could ultimately raise wages. Shorter hours would work against the tendency of the machine to create an oversupply of both goods and skilled workers. Without detriment to the basic interests of the employer, shorter hours would allow machinery to be introduced so that it would not displace large numbers of skilled coopers. Such a position conceded what the Cigar Makers' International Union refused to give up initially: skill was impossible to maintain before the inroads of technological innovation. The shorter day thus became the coopers' major instrument of adjustment.[39]

The Coopers' Union disappeared in the depression of the 1870's and was not revived until 1890. By that date, the industry was largely mechanized, but several steps in the assembly of barrels still required some degree of skill. The Union's reaction to the complete mechanization of the industry in the 1890's and the first decade of the twentieth century recapitulated much that had taken place during the period of mechanization that launched the original national union. Once again, the initial reaction was opposition. The union changed its constitution in 1897 by bar the union label to machine-made barrels; there was a boycott against the Pabst Brewing Company in 1899 because it switched to barrels produced in totally mechanized shops; and there were strikes by locals against the introduction of new machinery that would eliminate the last vestiges of skilled labor in favor of unskilled, nonunion men and children.[40]

As in the 1870's, it soon became apparent to the union leadership that such tactics could not succeed. Thus in 1899, the constitution was altered once again to allow the union label for machine work if specified hour and wage standards were met.[41] These standards were unrealistically high for the basically unskilled and semiskilled labor employed. Thus the constitutions of 1902 and 1904 permitted the organization of machine shops, such locals to make their own prices, subject to the approval of the general executive board of the international union.[42]

Thus the union accepted mechanization and sought to secure jobs as machine operators with the best possible wages and hours. In cooperage, opposition to machinery only led employers to reject any negotiations with a union that would not recognize their basic method of production. The hostility of many cigar makers to technological change had increased the proportion of nonunion workers in that industry, but the continuance of hand manufacture made such a development compatible with the interests of a significant group of skilled workers within the union. Lacking such a situation, opposition to machinery by the Coopers' International Union only forced men to leave the organization to find employment. The industry was now a mechanized one, and though a residue of hand work might persist, the Coopers' Union had to organize the machine workers, who now were the coopers. Any other policy was both ineffective and self-destructive.[43]

Both the cigar makers and coopers had hoped to conserve their skill and block the division of labor within their trades. The cigar makers' relative success in this endeavor was more a result of the conditions of the trade than any action by the union. In both industries, artisans had to concede their skill—the sole protection for their wages and working conditions—with no provision by employers or society to assist in the hard adjustment to technological change. Yet however destructive innovation might be, it also seemed inevitable in most industries. Thus skilled workers and their union leaders were in a constant tension between their realization of the dangers and the resulting impulse to fight change, and the realization that innovation could not be blocked and that efforts to do so would only make the effects of change all the worse. In trades such as cooperage, where employers viewed mechanization as an economic necessity, opposition was futile. The coopers came to this conclusion in the 1870's and once again in the early twentieth century, and in each case grudgingly made their peace with the machine.

## CHAPTER SIX

# LATER RESPONSES TO TECHNOLOGICAL CHANGE BY TRADE UNIONS

By the end of the nineteenth century, technological change had affected many of the skilled trades. Yet the passage of time and the extent of industrial change did not lead the trade unions to adopt a new policy toward innovation. At the end of the century, trade unions reacted much as they had in the late 1860's and early 1870's. Once again there was an initial impulse toward opposition, but most trade unions sought to restrain industrial change, usually through efforts to slow the pace of innovation and gain control over the new jobs for the members of the union. In this way, the new processes and machinery would be subjects for negotiation between the union and the employer.

Such a policy could take many specific forms. Thus machinists limited the number of machines that could be tended by a single worker, printing pressmen demanded more workers per machine than employers believed desirable, iron molders, clothing cutters, some building tradesmen, and glass workers limited the day's output, coal miners tried to set wages for machine work at levels that would limit the value of the innovation as a cost-cutting device, and printers blocked centralization of work in composing advertisements in favor of setting the same material separately at each newspaper. Most unions fought piecework, on grounds of the speedup, although the evidence suggests that restriction of output was clearly a major goal. These policies were designed to restrain innovation so as to lessen the negative effects upon the existing skilled labor force. They were manifestations of the labor movement's basic position: technological change must be controlled, and the interests of the existing skilled work force were one valid factor in justifying and carrying through such restraint.

75

Few trade unions sought to block the introduction of machinery into the industry. However, some influential locals of the Journeymen Stone Cutters' Association did seek to prevent the use of the planing machine, and they enlisted the national union in this effort. From 1895 to 1915, the planer was the central concern of the craft. The machine greatly reduced the time necessary to finish cut stone. Although hand methods were still needed for the final stages of the work, stonecutters feared that the demand for skilled artisans would drop as unskilled workers entered the trade to operate the planer. The effort by some locals to block the use of the machine in their areas divided unionized workers, encouraged the formation of an employers' association, and ultimately created a rival labor organization that accepted the employers' free use of machinery.[1]

Local unions of stonecutters initiated action on the planer by seeking to make its introduction into the trade less profitable. Restrictions on the number of hours that the machine might be worked and rules on the number of skilled union men who would have to operate it were designed to increase the cost of production with the planer and make it less attractive as an alternative to handwork.[2] These actions could be designed to slow the introduction of machinery and to ease the impact upon the skilled workers, and many trade unions used such rules as part of a policy of restraint. However, some of the stonecutters' locals insisted upon rules and wage rates that would prevent the use of the machine by making its operation so expensive and difficult as to leave handwork the more desirable alternative.

A key policy for those locals that sought to bar the machine was a prohibition against the shipment into their areas of stone that already had been worked by a planing machine elsewhere. The entrance of such stone from outside the area greatly hindered efforts by a local union to maintain hand operations. However, the attempt to restrict the movement of planed stone split the union, since stonecutters who worked in areas where the planer had been introduced, and in areas that shipped stone, saw no reason to stop work on stone intended for an area where the stonecutters' local still opposed the machine. By cooperating with their fellow workers elsewhere, they reduced their own work and income in order to oppose a machine that had already been introduced in their localities. Thus the failure to restrain the introduction of the machine in all areas created a divergence of interests between those who still worked by hand methods and those whose labor was already part of the machine process. Accordingly, in 1908, the Journeymen Stone Cutters' Association abandoned its efforts to set a national policy on the shipment of planed stone, after eight years of charge and countercharge within its ranks, and left any efforts in this area to individual local unions.[3]

Ultimately, the policy of opposition in some locals permitted the employers to weaken the existing union. "Independent" unions of stone-cutters were organized, committed to the acceptance of machinery. Employers often sponsored these independents or quickly negotiated settlements with them.[4]

Where markets were essentially local—as in the building trades—it was possible to restrain innovation in one area while it proceeded in another. As soon as it became clear that an innovation had been successfully introduced in some localities, trade unions usually adopted a policy of local option so that the national organization did not have to choose between those who believed they could successfully block mechanization and those who were obliged to acquiesce. The initial opposition to the planer by strong local unions of stonecutters was consistent with the actions of other unions, but the insistence on the policy of prohibiting the shipment of machine-planed stone clearly attempted to make those already at work on machines serve the interests of those who continued at handwork. This decision produced division and dual unionism. Its failure indicated the soundness of the local option policy followed by other unions that faced a mixed picture of hand and machine work.[5]

In reality, success in limiting the entrance of machine-planed stone into an area had depended upon the ability of the local union to convince employers that it was not in their interests to risk strikes (including sympathetic stoppages by other building trade unions) in order to use machine-planed stone from elsewhere.[6] Thus the decision by the Journeymen Stone Cutters' Association in 1908 only recognized the facts while resolving tension within the national union.

Where the planer was introduced, local unions insisted that skilled unionized stonecutters be employed on the machinery. This brought the skilled craftsman into direct conflict with the unskilled workers, who often had been used initially to operate the planer. Employers had turned to the unskilled because of opposition from a local union, or with the hope of eliminating the influence of the union in their business. Success in gaining the jobs on the planer for skilled unionized stonecutters varied with the area and the power of the local union, but many planermen did remain in the trade. Ultimately, the Journeymen Stone Cutters' Association accepted such men into the union, thus recognizing the permanence of the planer in the stonecutting trade.[7]

The stonecutters vigorously debated what action to take concerning the planer, and one is able to view with utmost clarity the positions that contended for dominance. There was little deception in discussions of a basic position toward the planer, although there was much evasion among branches of the union in carrying out national policy that conflicted with their basic interests. The continuance of open debate about how to respond to the planer reflects the disunity among stonecutters in their everyday reaction to machinery.

Three basic approaches to the planer were offered by the stonecutters. The first urged acceptance of the machine, with or without significant control over its operation by the union. The basic argument for such a policy was essentially negative: opposition could not work, and however distasteful technological change might be, it could not be blocked. Efforts to do so were useless and could weaken and divide the union. A Chicago stonecutter pointed out that the initial reaction of a majority of his fellow craftsmen in that city to the planer had been one of resigned hostility. It was useless to oppose machinery, for it was part of "the march of science which has been developed in this nineteenth century."[8] As a delegate to the convention of the Journeymen Stone Cutters' Association in 1902 put it: "I am opposed to planers and to the shipment of cut stone, and I am frank enough to admit that I know of no method of stopping it."[9] Strikes were the usual method of resistance, but they were an uncertain reed. Those who cautioned against opposition to the planer argued that strikes against machinery had been tried before and had failed. A strike policy would not only require unity among the stonecutters, which was unlikely, but support from the masons, who could also set stone and were demanding jurisdiction in the area. Beyond these problems of support within the ranks of labor stood the obstacle of the public attitude. ". . . To strike against or boycott the planers would at once enlist public opinion on their side, and with public opinion against us our cause would be lost from the start."[10]

Socialists within the ranks of the stonecutters maintained the traditional Marxist attitude toward machinery. Mechanization could not be blocked, and those crafts that had resisted in the past had been destroyed by the relentless advance of innovation. Machinery was not the real problem, for it merely allowed the production of greater wealth with less effort; instead the private ownership of machinery should be blamed for the evils resulting from technological change. The machine was here to stay; capitalism was not. "The class who use the machines must own them if they would escape the chains of slavery."[11] Thus the lessons of history and the reality of the moment convinced those who warned against opposition that any course other than acceptance was bound to fail.

A second group within the stonecutters accepted the ultimate victory of the planer but demanded that the union achieve firm control over the operation of the machines. Acceptance alone could place too heavy a burden of displacement and loss of bargaining power upon the stonecutters, yet opposition was bound to fail. In contrast, control of the planer by the union, through rules governing its use, combined with reductions in the work day, would allow stonecutters to "so control the machines that the displacement will be gradual and not sudden. . . ."[12]

One stonecutter believed a compromise could be reached with employers: the planer would do the rough work, with the final finishing of the stone still performed by hand.[13] More common was the belief that the union could convince the employers to use skilled workmen as machine operators, much as the typographers had become Linotype operators. Although the planer could be operated by an unskilled worker trained exclusively in its use, experienced stonecutters would be more efficient workers on the machine and would thus be preferred by employers. This policy would also protect the union by insuring that machine operators would be members in the same manner as in hand shops.[14]

Time was crucial in any such policy. Faced with opposition from the stonecutters, employers would turn to nonunion workers. These jobs would then be lost to the skilled stonecutters, especially as the unskilled became proficient in the use of the machine. If the skilled union men embraced the machine from the start, employers would be less tempted to turn to the unskilled, especially if in so doing they risked strikes. In practice, the refusal of some stonecutters to accept the planer did lead employers to hire unskilled men, who eventually had to be accepted into the union. By 1915, 10 percent of the jobs in the trade were held by planermen.[15] Those who had urged immediate adjustment to the planer, in an effort to control it, viewed these lost jobs as the minimum price of a policy of opposition.

Despite all such arguments for a positive reaction to the planer, the advocates of opposition remained powerful, especially in the years before 1910. In some of the stronger unions, such as those in St. Louis, New York, and Chicago, the use of the planer was delayed or severely restricted. The relative success of some encouraged others. Proponents of opposition pointed to the degradation of skill, the introduction of new workers who would create an oversupply of labor, the displacement of artisans, and the decline of wages.[16] Many stonecutters believed handwork was superior and that opposition should be based on the inferiority of machine work.[17] Some even suggested that the stonecutters organize their own firm, using the planer, and operate it to undercut employers who tried to introduce or continue use of the machine.[18] Opponents of the planer were convinced that concessions were ineffective: they would lead to the machine's control over the stonecutter and not vice versa.

Opposition appealed to those who viewed the planer as of benefit only to the employer, who would use it to undercut the gains made by stonecutters during the previous two decades. As the editor of the *Stone Cutters' Journal* put it in 1899: "We do not stand in the road of human progress, but we believe that labor-saving machinery should be used for the benefit of the people, as a whole, and not for the few, 'or not at all.'"[19] Lacking a social mechanism to secure the equitable distribution of the gains of technological change, and fearful that concessions by the union would strengthen the position of machinery and thus weaken the bargaining power of organized workers, the opponents of the planer accepted the "not at all."

The basic arguments for acceptance, control, and opposition were not unique to the stonecutters, but are the fundamental categories of response for American trade unions.[20] In varying combinations they have been present in most trades, regardless of the decade in which innovation threatened the existing labor force. Thus the debates within the cigar makers, coopers, and shoemakers in the late 1860's differ little in basic respects from the discussions within the stonecutters' union. Trade unions generally did not embrace a policy of opposition, and those that did usually abandoned such an approach after a short period. The decision was made in response to reality: most trade unions could not stop innovation, but they were more effective in restraining change in ways that protected the interests of the existing skilled labor force in an incomplete but still significant fashion.

A policy of restraint and control was certainly not a panacea, but merely the best choice available. Ultimately the precise form of response, and its success, varied with the conditions in a particular industry. These included the degree to which skill was still an asset in the most efficient operation of machinery, the relative strength of the trade union and the employers, and the extent to which handwork continued as an alternative to work as machine operators. Thus the control policies of the glass blowers and printers produced very different results.

The printers suffered a short-run displacement of one-third of the handwork force in the conversion to the Linotype, but the continued need for skilled workers made the hand compositor the best choice for Linotype operator, and the expansion of printing in the early twentieth century allowed skilled typographers to continue as machine operators under union rules. However, in glass bottle manufacturing, mechanization rapidly moved beyond semiautomatic machinery that could still use the skilled worker with profit. The existence of powerful nonunion employers, plus a jurisdictional struggle between the Glass Bottle Blowers and the Flint Glass Workers, added to the problems of adjustment and control. Unlike the printers, few glass blowers became machine workers, even when semiautomatic machinery was still used, in comparison with those who managed to continue in the trade by shifting to specialized forms of work that still used hand methods. As the industry converted to automatic machinery in the years 1905–1915, displacement of skilled workers was widespread since the new machinery did not need them in any capacity.[21] In essence, the bottle blowers carried through a policy of delay: it reduced the impact of innovation upon the skilled work force to the extent possible but failed to create conditions that would integrate the skilled workers into the new productive process under union rules that would maintain their interests.

Technological change also influenced the relationship between the national officers of a trade union and the local leadership. The national leaders generally called for a policy of control. They viewed the introduction of machinery from a nationwide perspective and considered its impact upon the continued existence of the union as well as the threat to those facing displacement. Local leaders and many rank and file members more often demanded a policy that would exclude the machine from their particular area, even if such a policy seemed unrealistic from a national or long-term viewpoint. The iron molders' union illustrates the division very clearly.

Initially, mechanization elicited strong opposition from the rank and file, but acceptance from the national leadership. Skilled molders refused to operate the machines, or, if employed, they operated them at a level far below maximum efficiency. These actions by the skilled workers led employers, who wanted to mechanize, to turn to unskilled labor, thus removing the machines from the control of the union.

The leadership of the Iron Molders' International Union fought this development, arguing against opposition to the machine on the grounds that it had failed in other trades and thus could not be expected to succeed in iron molding. Acceptance allowed the union to bargain about the conditions under which machinery would be operated. In addition, a continued union presence might secure regulations that would maintain hand methods to the extent possible.

After years of debate within the union, the leadership prevailed, and a policy of control was adopted. Machine operators were admitted into the union in 1907, and skilled molders were encouraged to work machines—where they had been introduced—at rates negotiated by the union.[22]

The call for opposition was so strong in the iron molding trade because the rate of introduction of machinery was exceedingly slow. Machinery had been introduced in the 1890's, but by 1916 only 25 percent of the foundries had installed it. Thus relatively few of the skilled molders in the union actually were faced by the threat of machinery, and still fewer had accepted employment as machine operators. With handwork still so plentiful, it was relatively easy for skilled workers to avoid machine shops, and many did so.

In contrast, printers faced the immediate spread of the Linotype, with rapid and significant displacement of handworkers. The International Typographical Union moved to a control policy without significant opposition from within the rank and file. It was widely recognized that the Linotype could not be blocked, that it was spreading quickly, particularly in the newspaper branch of the trade, and that skilled handworkers could become the machine operators, if a policy of control were instituted before any large group of less skilled men entered the trade as Linotype operators. In fact, the International Typographical Union even used union men to break a strike by one of its own locals against the introduction of the Linotype.[23]

As we have pointed out, the cigar makers' union faced a division much like that of the iron molders, with the national leadership calling for greater acceptance of innovation. Again skilled workers maintained opposition because opportunities for handwork (in this case in the manufacture of higher-priced cigars) still allowed many union members to avoid a confrontation with the machine.[24]

The complex relationship between opposition and control also can be seen in the glass industry. Within the period 1898-1905, semiautomatic machinery rapidly replaced hand methods in the production of lamp chimneys and wide-mouthed glassware—two major elements in the industry. The machinery still required skilled workers for its most effective operation, and the two unions of glass workers involved—the Flint Glass Workers and the Glass Bottle Blowers—adopted a similar policy of control. However, a jurisdictional dispute over whose members should operate the new machines handicapped the efforts of both to secure the jobs as machine operators.[25] The frank discussion of the machinery issue in both unions clearly reveals the considerations that produced the ultimate policy of control.

The glass workers would have preferred to block the machine, if that were possible, and the officers of the American Flint Glass Workers' Union even explored ways to do so. One possibility was to join those employers who opposed the introduction of machinery, because of its competitive effect upon them, in a joint effort to block the innovation.

Despite the popular image of the late nineteenth-century businessman as an innovator, employers actually responded in various ways. Some welcomed attempts to restrict output to the level that was most profitable for the industry. Others sought to mechanize rapidly, despite opposition from other employers and the existing work force. The attitude of an employer toward machinery was motivated by many factors, including the impact of the innovation upon the competitive position of the firm, the financial resources of the company, the strength of any opposition from the existing work force, and the relative profitability of the innovation when compared to established methods of production.[26]

In the carpet industry, hand methods continued long after the introduction of more efficient power looms, because of the high initial cost of the new machinery. The continuing royalty per yard that had to be paid to the patent owner, and the relatively poor performance of the power looms with cheaper quality goods (weaker yarns were likely to break, stopping the machine and reducing its competitive edge over hand methods) also slowed the introduction of the machinery.[27] In baking, mechanical equipment was available in the late nineteenth century. However, in 1899, 78 percent of the bakeries were still small operations with fewer than four employees. Less than 10 percent used any power equipment. The baker was a small employer who resisted mechanical change, not only because of his customs and craft tradition, but because the machinery was designed for large-scale operations beyond his means.[28] Silk companies opposed the introduction of rayon, and brick manufacturers argued that concrete was unsafe.[29] Large companies also resisted technological change to protect their position. United States Steel was slow to innovate in the early twentieth century because of its large investment in older equipment. From 1896 to 1911, General Electric and Westinghouse agreed not to patent innovations that would hurt each other, and thus superior light bulbs were kept off the market.[30]

It would be quite incorrect to argue that labor unions have opposed innovation while businessmen have favored it. Many employers—large and small—also opposed innovation, either at the time of the first significant departure from hand methods, or after an earlier acceptance of technology had given a firm an excellent competitive position as well as a plant that would be expensive to replace. As one observer has put it: "In actual fact, resistance by labor has not even approximated resistance motivated by business interests. Labor is neither organized well enough nor financially able to be effective."[31] Thus there was ample precedent for employers opposing innovation, but usually not in combination with a trade union.

In the late 1890's, the Flint Glass Workers' Union and a group of employers considered the possibility of buying the patent for the Owens lamp chimney machine—the most successful one in operation—in order to block its further use. The ownership of the patent not only would protect hand methods from technological change, but also would allow the union and the cooperating employers to use the machine selectively against other manufacturers who tried to operate nonunion shops.[32] Thus the machine could be a weapon to impede competition among manufacturers resulting from reductions in the cost of labor.

The plan failed for three basic reasons. First, the price asked for the patent was beyond the means of the union and the cooperating employers, although the union proposed in 1899 that this could be overcome by advancing the selling price of lamp chimneys to gain the needed funds.[33] The control of the patent would make it likely that such higher prices could not be forced down. Second, the employers were not united in their attitude toward machinery, and some would seek to mechanize whatever the action taken by the union and the cooperating group of employers. Many of these manufacturers operated nonunion shops and were beyond the influence of the Flint Glass Workers' Union. Third, the opportunity for manufacturers to mechanize was still a real one, even were the Owens machine to be purchased, since at least eight other machines were under development. Their use would certainly increase should the Owens machine disappear from the market.[34]

The Flint Glass Workers' Union also discussed cooperatives and profit sharing as means of meeting the threat from machinery. The cooperative shop would continue hand methods and would guarantee employment to all members. Of course it would have to face the competition of machine shops. However, "Economically managed the plant will be in a position to hold its own against any competition because if it becomes necessary to make any sacrifices, the profits can be sacrificed to save the wages, whereas if working for capital, wages would be sacrificed to save profits."[35] The scheme does not seem to have gained many supporters. Certainly the experience of cooperative cooperage shops in Minneapolis should have been a warning to potential cooperators among the glass blowers. The coopers had been unable to meet the competition of mechanization except by introducing machinery themselves, with the consequent displacement of skilled members.[36]

For those who lacked the capital to enter a cooperative scheme, President W. J. Smith of the Flint Glass Workers suggested a profit and loss sharing plan. This would appeal to employers who also felt threatened by machinery, since they would share losses with their workers. On the other hand, the sharing of the profits would help cushion the worker against the impact of the machine upon his wages.[37] Such a plan would be likely to appeal to employers, who could not expect any profits in the face of competition from machinery, but their workers would not long accept an arrangement that provided losses rather than profits.

The Glass Bottle Blowers' Association tried to compete with the semiautomatic machinery for a brief period in the late 1890's by reducing the piece rates for handwork. The objective was to make it unnecessary for employers to resort to machinery in order to meet the competition from those manufacturers who had already abandoned the older hand methods. The policy failed, since the reductions were sharp enough so that many skilled workers sought out the shops that still used hand methods for types of glassware that could not be made on the Owens semiautomatic machinery, and where piece rates had not been reduced. In addition, many employers concluded that despite the cuts in piece rates, the machine would still be more profitable.[38]

Ultimately both unions of glass blowers adopted the control policy. As President W. J. Smith of the Flint Glass Workers' Union remarked: "to work machinery will throw some of our members out of employment, but to

refuse to work machinery will throw more members out of employment. . .
for if the workmen in the association will not work machinery, those outside the association will work it. . . ."[39] The appearance of automatic bottle-
making machinery in 1905 undermined the policy of control since the new
machinery no longer required the services of skilled glass workers for its
most efficient operation. As a result, the displacement of skilled workers
soared. Thus despite the strong unions in the trade, skilled glass blowers were
left largely to the vagaries of the uneven effect of machinery in different
portions of the industry or in different localities.

For the machinists, the issue was not machinery itself, but the manner
in which it was to be operated. In the first decade of the twentieth century,
the International Association of Machinists was split over the issue of piecework. The Association opposed piece rates, and a number of locals fought
bitter strikes to prevent their introduction in place of day wages.[40] President James O'Connell shared the hostility to piece rates, but he headed the
group within the union that believed opposition was futile because of the
determination of the employers to use piece rates as a method of increasing
the output of their machinery. In addition one had to consider the willingness of many machinists to accept the new wage system despite the opposition of the national union.

Opponents of piece rates—or an alternative premium plan that would
have shared all gains above an established average output equally between
employer and employee—believed the ultimate result of these devices would
be a reduction of the work force. The basic argument flowed from the overproduction thesis. Industry was unable to dispose of the existing product
in any consistent and reliable way. If productivity per man increased with
the introduction of piece rates, the disposable production would be made
in a shorter period, and employers would then lay off workers. Fewer men
making more goods meant unemployment for some of those previously at
work. These idle men then became a club the employers might use to beat
down the wages of those still at work. As a proponent of increased productivity put it, the machinists' objection was "to everything that tends to min-
imize labor."[41] The argument against piece rates ignored the contention
that consumption and employment would increase with improved productivity and a lower cost of production.[42] As we have observed, trade unionists were skeptical of the increasing consumption argument, since experience
revealed the distressing disparity between output and consumption which
was basic to the overproduction thesis.

Agreement among machinists on the danger of piecework failed to produce unity on how to confront this threat. As with the stonecutters, the
machinists divided because so many of the union's members accepted the
innovation. These workers regarded the attempt of the International

Association to block piecework as a direct threat to their employment. President O'Connell's own willingness to accept piece rates was predicated on the belief that when faced with a conflict between union rules against piece rates and a job that was offered on a piece rate basis, many members would reject the association's restrictions. To fight against piece rates under these conditions was to weaken the union.[43]

As piecework continued to spread among machinists, the convention of 1903 sought to reaffirm the traditional hostility of the International Association to piecework and the operation of more than one machine by a machinist. It submitted to referendum a resolution demanding enforcement of the union's rules against piecework and the operation of more than one machine. O'Connell regarded such a stance as unrealistic, and the referendum sustained his judgment by a lopsided margin.[44] Local strikes against the introduction of piecework were frequent, but generally were lost.

Efforts by trade unions to restrain technological change were also undercut by the existence of nonunion shops. In the early twentieth century, hosiery weavers in Philadelphia were able to enforce a rule against the operation of more than one machine because of their tight control of the specialized work force needed in this trade. However, the union had to give up the restrictions because of its failure to organize competing mills in areas outside the city.[45] As a union upholstery weaver put it, in commenting on the conditions of his trade in the highly organized shops of Philadelphia, workers could secure "a high wage scale and good working conditions, with no work."[46]

Coal miners also faced the threat of mechanization in the decades following 1880. They did not respond by opposition to the introduction of machinery, nor simply by a policy of control of the jobs as machine operators. Instead, the miners' organizations sought to limit the difference between the cost of a ton of coal mined by hand and a ton of coal mined by machinery, to the smallest figure possible, in order to allow hand-cut coal to remain competitive, and to slow the rate of mechanization by reducing the potential profit.

Coal mining machines had been discussed as early as the 1860's, but they were first introduced about 1880 into the bituminous fields.[47] The machines were basically power cutting tools that could be used most efficiently in thick seams of coal. Despite this limitation, 11 percent of the coal produced in Illinois was machine cut by 1890. After 1890 the United Mine Workers reported a steady increase in the percentage of the total national production mined by machine, rising from only 6 percent in 1891 to 35 percent in 1906.[48] Although some miners in the 1880's believed the machine would fail, it was clear by 1890 that such would not be the case. The miners had to assess the effects of the machine and develop a policy concerning innovation.

The coal cutting machine did not displace the hand miner so much as convert him from a skilled worker—who did most of the mining operations, such as blasting, cutting and timbering himself—to a specialist who performed only one part of the process. This division of labor was as much responsible for the increased output in mines that used machinery as the mechanical device itself. Division of labor enabled the employer "to concentrate the work of the mine at given points, and it admits of the graduation of wages to specific work, and the payment of wages by the day."[49] The machine did not eliminate skill, but its use meant that a smaller proportion of the total work force had to be recruited from the ranks of more highly paid skilled labor.

A study in 1890 revealed that in one Illinois mine, which used both hand and machine processes, handworkers earned an average of $2.01 per day. Operators of the coal cutting machine earned $2.41 per day, blasters earned $2.30 per day, and timber men earned $2.12 per day. These specialized workers performed all the skilled work formerly done by a hand miner where there was no division of labor, but they numbered only 49 as compared with the 174 laborers earning $1.73 per day and the 38 helpers who were paid $1.75 per day.[50] Thus a large proportion of the total work force was paid less than the average daily wage of the hand miner.

It was significant that some men would earn more in the machine process. If the rate of introduction of machinery were not overly rapid, many of the hand miners would not be greatly hurt by the change. Some would become cutters, blasters, and timber men, and they would continue to earn wages comparable to or higher than hand miners. Since coal mining was an expanding industry until the 1920's, the need for skilled specialists continued, even with the increase in strip mining which employed entirely new methods.[51] Other miners dropped from the trade by attrition—which not only included retirement and disgust with conditions in the mines that led young men to leave the mining districts—but injury and death from the many mining accidents. The unskilled positions were filled by newcomers from the ranks of native and foreign-born migrants into the mining areas. Thus the key for the hand miner was the rate of technological change, and his reaction to machinery reflects an understanding of this basic point.

The initial response of miners in Illinois was typical of workmen in the earliest stage of mechanization. Some counseled acceptance; others called for opposition, either through boycotts of the machine or by a reduction in the rates for hand mining in an effort to undercut the machine and drive it from the fields. Such opposition had popular support in the St. Clair mining district, where physicians "pronounced the effect of the compressed air upon machine operators to be injurious to health."[52] Such an opinion is more a comment about the physicians' antipathy to machines that might threaten the social and economic structure of mining towns than a medical judgment of much use of miners, whose every working day was injurious to health.[53]

By 1886, the National Federation of Miners, holding its second convention, proposed the policy that was to be the basis of later efforts of the United Mine Workers to respond to the machine. Opposing strikes against mechanization, the convention resolved that machine miners "be paid a price to equalize wages with pick miners."[54] In 1891, the newly formed United Mine Workers set the policy of an established differential between the piece rate paid for pick mining and the piece rate paid for coal mined with the help of machinery.[55] This price varied from area to area, reflecting differences in the amount of coal that could be reasonably expected in a day's labor and the strength of the union in a particular locality.[56]

The United Mine Workers rejected the possibility of preventing the introduction of machinery. President J. P. Jones of District 6 (Ohio) expressed this position at the state convention of 1892.

The tendency of the age is to substitute machinery for hand. This is noticeably so in our business, and every day machines are increasing in numbers and improving in efficiency. The displacement of hand labor by this method has only begun. It is matterless if we approve or disapprove. My judgment is that mining machinery is but in its infancy, and hence we should prepare ourselves to yield with as much grace as possible to the inevitable, and turn our attention to securing the best terms possible.[57]

President John Mitchell stated the position of the United Mine Workers in 1900 when he argued that the Union "does not take the untenable position that the introduction of labor-saving machinery in coal mines should be discontinued, but we cannot be accused of retrogression by demanding equal opportunities to compete in the markets in both hand and machine mines."[58]

As developed in subsequent years by the United Mine Workers, the policy aimed at a fixed differential of seven cents per ton for machine-mined coal.[59] This meant that the union would permit only seven cents less per ton for coal mined by machine than for coal mined by hand. Such a rate would allow machine coal cutters to do as well as or better than hand miners because their production per day was greater. Thus machine operators would continue to support the union, since its policy on machines did not hurt their interests. The divisions that developed among the stonecutters and machinists were avoided.

Yet the policy was basically designed to serve the interests of hand miners, and it did so in two basic respects. By maintaining a fixed differential of only seven cents per ton, and recognizing the comparative weakness of the machine in many coal mining situations, the hand miner would be able to compete with the machine in some types of mining. Even where this was not possible, the fixed differential would prevent the massive profits that accelerated the introduction of machinery. Since the miners never planned to block innovation, a slowing of the rate of change was the basic objective. It would cushion the effect of the machinery upon the miner, but still allow mine owners, who had optimum conditions for use of the machine, to introduce the new devices without interference from the union.[60] In addition, the United Mine Workers demanded the reduction of hours to increase the possibilities for absorption of those miners who were displaced by machinery.[61] It was a well-designed policy that sought to avoid the pitfalls of opposition, that engendered conflict within a union and hostility from employers, or simple acceptance, which would have done little to cushion the impact of technological change upon the hand miner.

The United Mine Workers made determined efforts to establish the seven-cent differential and the shorter work day in its negotiations with the operators. But success was minimal. Many operators opposed the seven-cent differential precisely for the reason that it appealed to the union: the rate was set so that only mines that had optimum conditions for the use of machinery would profit from the change to the newer process. It was thus viewed as an effort to impede mechanization in order to maintain hand methods.[62] Where hand mining was cheap, particularly in nonunion areas, owners were less attracted to machinery. Operators who maintained union agreements pointed out that mechanization could be the means by which they were able to meet the United Mine Workers' wage scales and conditions of employment yet still compete with the nonunion mines.[63] Thus the failure of a union to organize throughout its trade once again acted as a powerful barrier to efforts to meet the threat of mechanization.

The owners in Illinois, where coal mining machinery was well suited for many thick-seamed bituminous mines, and where the United Mine Workers had a strong organization, agreed to the seven-cent differential; but owners in other states balked, and the union ultimately accepted a compromise which set the price for machine-mined coal as a percentage of the existing price for hand-cut coal in unionized mines. In 1900, a 75 percent formula was used in Indiana, which meant that machine mines paid eighteen to twenty-one cents less per ton than hand mines.[64] This figure was more likely to encourage mechanization than the flat seven-cent differential demanded by the United Mine Workers. In addition, the three-quarter formula meant that if the price paid by the owners for hand-cut coal increased, the spread between the hand and machine price became wider. Thus gains in wages for unionized hand miners would encourage mechanization. The United Mine Workers sought to establish a fixed differential, even if greater than seven cents per ton, but did not win the point in Ohio until 1914 and in western Pennsylvania until 1916.[65] The inclusion of the machine-versus-hand price issue in collective negotiations with the owners was a gain for the union. However, the relative weakness of the United Mine Workers made it difficult to win the seven-cent differential or the shorter day in areas where employers were opposed.

The United Mine Workers never claimed that its policy was meant to block mechanization, and clearly it did not do so. Even though the seven-cent differential prevailed in Illinois, the percentage of the state's total coal production mined with machinery increased from 18 percent in 1898 to 56 percent in 1914 to 65 percent in 1921. Moreover, 30 percent of the total production was strip-mined. Thus only 5 percent was still mined by hand. The figures for Indiana and Ohio, where the differential had always been considerably higher than seven cents, show the same result. In the early 1920's, only 7 percent of the coal mined in Indiana was obtained by hand methods; 53 percent was mined with machinery, and 40 percent strip-mined. In Ohio, 90 percent of the coal was machine mined; only 3 percent was mined by hand, with the remaining 7 percent produced in strip mines.[66] The differential was reduced in most mining areas between 1912 and 1921 without any effect upon the rate of mechanization.[67] In fact, hand mining was most extensive in nonunion areas, where lower wages made the installation of machinery less attractive to owners. Nationwide, coal produced by machine methods had increased from 25 percent in 1900 to 33 percent in 1905 to 42 percent in 1910 to 55 percent in 1915 to 60 percent in 1919.[68]

Union officials had foreseen this development as early as the late 1880's, and the policy of the differential was used to cushion the effect of mechanization, to the extent possible, while avoiding divisions within the union that could weaken its overall effectiveness. The policy also recognized the futility of trying to block mechanization in unionized mines when machinery offered owners, who had union contracts, a means of meeting competition from the many nonunion mines. Considering the strength of the union and the economic conditions of the industry in the period from 1890 to 1920, it was a well-conceived course of action.

Even in situations where labor unions did not seek to restrain innovation, employers might choose to provoke a conflict over the distribution of the gains from improved technology. This was certainly the case in the Homestead strike of 1892. The steel industry in America had been marked by sharp competition since the 1860's. Pools were suggested as a method for reducing competition in the industry, but Andrew Carnegie and a number of other entrepreneurs rejected this approach. Instead, they viewed technological advances, combined with the competitive situation in the industry, as a means of securing dominance. Between 1870 and 1890, many technological advances were introduced that increased productivity and lowered costs. Those who innovated secured a competitive advantage. In the process, the skilled workers were reduced in number to a small proportion of the work force, and with the conversion of jobs from skilled labor to semiskilled machine tending, wages fell. The combination of increasing productivity per man and falling wages dramatically lowered costs per ton of steel.[69]

The response of the Amalgamated Association of Iron and Steel Workers, which represented the skilled work force in the steel mills, was to accept mechanization. As President William Weihe testified following the Homestead strike: "The organization never objects to improvements and makes allowances in every particular where there are improvements, and in steel mills there are special clauses, and wherever there is an improvement made by which certain men will be done away with, then their jobs will be done away with. There is no objection."[70] The association accepted lower piece rates where production clearly increased through mechanization, although it did try to set limits on output to maintain work for as many men as possible.[71] These efforts were not overly successful, in part because employers strongly fought any barriers to productivity gains, and in part because the association's own members, under pressure from management and lured by higher wages, were often willing to exceed the limits.[72]

Henry Frick took command of the Homestead Steel Works in 1889, and he emphasized the reduction of costs through increases in productivity. Carnegie had been friendly toward the association in the 1870's and 1880's, when skill was still vital in steel making; but by the early 1890's, Frick was prepared to crush the union, which had become much less important with the continued mechanization of steel making.[73] He demanded reductions in the piece rates paid skilled workers, claiming that technological improvements had increased production, which would allow the men to earn as much with the lower rates.

The association responded that greater productivity was more the result of harder work by the men and the switch from twelve- to eight-hour shifts than any new technological gains. Moreover, the union argued that the continuing decline in skilled jobs amply repaid the company's investment in machinery.[74] Professor Edward Bemis was quite aware, at the time, that the major issue in the Homestead strike was "who should gain by the improvements, capitalists or wage-worker, in a business where, through combination, patents, secret processes and the tariff, the consumer might expect some time to elapse before securing all of his share in the benefits of increased efficiency."[75]

Frick was intent on gaining all possible shares. Despite the willingness of the association to accede to the displacement of its members by technological advances, he was prepared to destroy the union in order to give the company maximum freedom to retain as large a portion of the gains in productivity as possible. The dominance of Carnegie Steel had been built on technological leadership, as translated into competitive advantage, and its ultimate control of the industry, which was achieved through the creation of United States Steel in 1901, was dependent on extending this technological lead. The association was a potential barrier and thus had to be eliminated. Neither violence, including his near assassination, nor the intervention of politicians, could sway Frick from his purpose. Ultimately the association was defeated, and effective unionism was driven from the steel industry until the 1930's.[76]

In the steel industry, the ultimate destruction of the union of skilled workers was a by-product of technological gains. In other situations, employers turned to machinery for the primary purpose of weakening or destroying unions. The McCormick Harvester Company introduced machinery in 1886 with the primary purpose of eliminating the skilled molders and breaking the influence of their union. The company succeeded in this objective, relying upon its large profits from the 1879–1884 period as a financial cushion. It was also aided by police assistance, support from other employers, and the hostility of the press to the strike that resulted from the company's actions.[77] A major reason for the introduction of power looms in the carpet industry in the 1840's was to weaken the position of the skilled weavers and thus reduce strikes.[78]

In the cutlery industry of the Connecticut Valley, skilled workmen continued to be important throughout the nineteenth century. They were strongly organized and had won closed shop agreements, regulation of apprentices, and jurisdiction over the existing machinery. The employers thus sought new technology not only for the productivity gains, but "to get the best of these devils." After the introduction of improved machinery, the Meriden Cutlery Company claimed that "our men are as meek as kittens." The power of the unions was undercut, and between 1906 and 1912 the cutlery trade in the Connecticut Valley became an open shop industry.[79]

The introduction of machinery to last shoes threatened the remaining group of skilled handworkmen in the shoe industry in the 1890's. Many employers made no secret of the fact that they would use machinery to force concessions from the lasters' unions or break their agreements and operate as open shops with semiskilled workmen operating the new equipment.[80] Since the lasting machinery on the market in the 1890's still could be worked most efficiently by skilled workers, much as the Linotype could be operated to advantage by a skilled printer, employers generally maintained their skilled lasters if the workers would accept the machine. The threats were usually to gain a bargaining advantage, not to destroy the union but as the machinery was improved, in the twentieth century, the skilled lasters disappeared, just as so many craftsmen had before them.

Whatever the precise figures for jobs created by machinery, as against those destroyed, Eugene Debs spoke for the labor movement in insisting that the result for the present generation was that many workers were in enforced idleness—"the most serious menace that confronts our institutions and civilization."[81] Arguments of progress and national interest meant little to those intimately affected by technological change. Most skilled workers in the late nineteenth century agreed with the craftsman from Ohio who contended that "Labor saving machinery has made some men richer, most things cheaper, and the working classes poorer."[82]

Organized workers attempted to restrain innovation, by a wide variety of restrictions and rules, in order to maximize work and thus improve their bargaining position. These regulations had the additional purpose of establishing working conditions that would make the introduction of machinery less profitable, which might slow the rate of change. Machinery was only one more weapon of the employer in the ceaseless competition between labor and capital for advantage.[83] Yet it also was a major element in the competition among capitalists themselves, which forced employers to use machinery, child labor, botch workers, and any other expedient, in an effort to gain an advantage over their rivals.[84] The labor movement represented one of the few institutions dedicated to noncompetitive arrangements in the midst of a competitive economic system.

A major reason for the development of trusts in the late nineteenth century was to offset the negative effects of industrial change upon business, as reflected in overproduction and intense competition. Trade unions sought the same general objective: control over technological change so that it could be made compatible with their interests. The trust and the labor union demanded restraint of a freely operating capitalism that could destroy their interests. It is thus not surprising that those Progressives who emphasized trust-busting were often unsympathetic or openly hostile to the restrictive practices of trade unions, including the closed shop and restraints upon output.[85]

In a society that lacked a philosophy of governmental action that would sanction adjustments between the dictates of competitive capitalism and the interests of those who might suffer from the untrammeled development of the economy, labor unions depended mainly upon their work rules and agreements with employers to try to weaken the blow when it reached the level of the worker. Aside from the restriction of immigration and regulation of the labor of women and children, the weapons used by trade unions were nongovernmental. In the 1880's and 1890's, even eight hours was to be achieved by the economic action of trade unions.

The success of such a policy obviously depended upon the strength of the union, but machinery worked to undermine the skill upon which the power of trade unions was based. Innovation also created a massive new body of workers dependent upon machinery for their jobs and thus little concerned that some other group of workers might lose their employment in the same process. Industrial progress provided real benefits for those not directly challenged by displacement, and it cut off much public support from the trade unionists. Thus skilled workers found that they could not challenge the machine without attacking the substantial interests created by it in all classes of society.

Efforts to restrain mechanization in a trade ultimately succeeded or failed because of basic economic conditions in the industry, over which the worker and his union had little control. Thus in many of the building trades, innovation left considerable skill intact, which allowed unions to regulate the supply of workers. It then became a question of the strength of the union and its ability to control the skilled work force in an industry of small contractors, seasonal demand for the services of employers as well as workers, and local markets. The unions in the building trades recognized these factors and used them to advantage; they did not create them.[86]

Failure to develop a strong welfare state left no mechanism by which divergent interests might be reconciled with the needs of a larger public. Even within an industry, employer and worker rarely bargained collectively, because of the refusal of businessmen to accept any restraints on their freedom of action. There was no respite from the clash of interests, no means of adjudication beyond the power of the stronger. Trade unions understood this perfectly, and they used whatever means were available to safeguard their interests to the extent possible. Since society did nothing to cushion the effects of innovation for the individual worker, unions felt that they had to impose restraints upon industrial change. The trade union was ill-suited for such a task, but labor leaders clearly made the choice to protect the concrete interests of their members and not to serve some vague general good. Perhaps the real tragedy of America's industrial revolution was that labor unions had to face such a choice at all.

## CHAPTER SEVEN

# APPRENTICESHIP AND
# INDUSTRIAL CHANGE

Even before the burst of mechanization in the 1880's, division of labor and the shift of production into factories were replacing older methods based upon skilled hand labor in small workshops. The changes in the three decades prior to the 1880's primarily involved dividing skilled tasks in an effort to increase production. Employers realized that a diminution of the skill required not only could increase production through a division of labor, but also could reduce the bargaining power of the skilled worker. Accordingly, employers sought to increase the number of apprentices and helpers so as to provide the necessary pool of labor. Even without extensive mechanization, the skilled worker opposed this development as a threat to his basic interests, which he believed centered on his control over the supply of labor. Much of the labor movement's response to industrial change was an elaboration of this basic principle.[1]

Labor leaders were quite clear on the importance of limiting the number of apprentices as a means toward the broader objective of control over the supply of labor within a trade. The addition of a large number of apprentices to the trade reduced wages in several ways. First, most of the apprentices were poorly trained, since employers were interested in using them in limited, divided tasks and little concerned with their general knowledge of the craft. The result was that these boys became the so-called botch workmen, who could not produce the quality or quantity of work that tradition or union rules established. Yet they were available to compete with skilled men especially where division of labor had been introduced. The result was downward pressure on the wages of the skilled toward that offered to the less skilled.[2]

Even more important, the increase in the number of workers in the trade was not limited by the employment available. In prosperous periods, all workers might be employed, but this would not be the case in slack times. Employers would then seek to cut costs by reducing the work force in number and reducing wages as well. The result was not only unemployment for some of the skilled workers, but once again pressure on those still at work to accept lower wages or be replaced by younger workers.

A demand by employers to modify the existing limit on the number of apprentices prompted a strike in 1864 by the St. Louis local of the Iron Molders' International Union. The union argued that "If the number of journeymen for one apprentice was too great we were willing to reduce it, or if the employer or Government will guarantee to keep employed all the men . . . even if they reduce the number of hours of labor, again we acquiesce. But we too well know the program in slack times—all the journeymen off, and only the boys kept at work."[3] Labor leaders viewed the insistence of employers on a greater number of apprentices as a clear effort to glut the market, which would reduce the bargaining position of the worker, whether he was represented by a union or not. An increased number of workers would create underemployment, increase the severity of seasonal or cyclical unemployment, and create a pool of men to use as strike breakers, thus further weakening the bargaining position of skilled craftsmen.[4]

The issue was so serious a threat to skilled workers that it had figured prominently in the formation of trade unions in the 1850's.[5] The printers sought to regulate the number of apprentices from the inception of a national union in 1850. The convention of that year issued an "Address to the Journeyman Printers of the United States" which stressed the "perpetual antagonism" between labor and capital. "Every addition to the number of laborers in the market decreases their power; while the power of capital grows in a ratio commensurate with the increase of capital itself." Thus a union was necessary to defend the interests of the worker because it "destroys competition in the labor market, unites the working people, and produces a sort of equilibrium in the power of the conflicting parties." Limiting the number of apprentices was necessary to control the number of workmen and prevent the employment of "herds of half men at half wages."[6]

It proved a difficult objective to achieve. The convention of 1850 called for a five-year indenture system, but opposition from employers and resistance from locals (who opposed a rigid national rule that might force them into conflict with employers) led to an inconclusive policy that vacillated between the establishment of a five-year rule and a system of local option. Disparities between locals continued concerning the number of apprentices permitted and the length of the apprenticeship.[7]

The labor movement rejected the charge that its hostility to unlimited numbers of apprentices proceeded from a desire to monopolize the jobs in the trade for the selfish interests of those already employed.[8] Martin Foran, President of the Coopers' International Union, argued strongly that limits on apprentices were "to prevent the disastrous effects of the competition sure to follow should the number of craftsmen in the trade greatly exceed the demand." Such rules were necessary to prevent unemployment, and this served the entire community by eliminating the demoralization that resulted from idleness.[9] The objective was not monopoly, but rules that would "regulate the supply of labor so as not to exceed the demand."[10] The division of labor made such control all the more difficult and thus intensified competition among workers in a trade to their mutual disadvantage. Moreover, by reducing the importance of skill, it threatened to open the trade to "the whole mass of the unemployed."[11]

It is thus not surprising that William Sylvis attacked the division of labor as an instrument devised by capital to deprive the mechanic of his independence. Sylvis quickly added that he was opposed not to a division of labor itself but to its abuses—a statement that would later be repeated many times with regard to machinery.[12] Clearly apprenticeship rules had become a device for offsetting the effects of changes in the system of production upon the established work force, rather than a method for training new craftsmen.

Despite the importance of controlling apprentices for the trade unions of the 1850's and 1860's, only minimal success was achieved. Employers frequently violated union rules on the number of apprentices, and this led to many local strikes; the decentralized nature of many trade unions made it difficult to enforce uniform regulations, which produced wide disparities within union shops; workers themselves often ignored the regulations, since they desired to have their own children or those of relatives and friends apprenticed in the trade; labor unions often found it difficult in good times to provide a full complement of union journeymen, which led to claims that there was a shortage of skilled workmen, necessitating an increased number of apprentices; unions had little control over the division of work, which provided a constant inducement to employers to press for a greater number of apprentices or helpers; and the large number of nonunion shops, where apprenticeship was uncontrolled, led to constant pressure on unions to modify their regulations.

This situation led some labor leaders to press for governmental action. There were demands for state laws that would reinstate the dying system of indentures. A committee reported to the national convention of the Plasterers' Union that such a law would "aid in forcing back to the agricultural, mining, and manufacturing districts the surplus population that centers around the mechanical branches in large cities."[13] The New York Working-men's Assembly demanded a law that would set the period of indenture as three to five years and would place legal responsibility for insuring that apprentices learned the trade fully upon the employers. Violators would be fined, and apprentices who broke their indenture would be subject to arrest, trial, and confinement in the "House of Refuge or Correction until twenty-one years of age."[14] There were also suggestions for state examinations to certify that craftsmen were qualified in their work, in the same way that state boards licensed lawyers, doctors, or engineers. Such licensing not only established quality standards, but kept the number of men in the profession limited.[15] These efforts to involve the government in the apprenticeship issue produced meager results, and the problem of controlling the number of workers in a trade continued.

The machine was not a major threat to most skilled craftsmen in the 1860's, but it did affect shoemaking, contributing to the formation of the Knights of St. Crispin in 1867. The Crispins' response to the threat of ma-chinery was basically an effort to control the supply of labor through ap-prenticeship regulations. Thus it differed little from the response of other trade unions to division of labor, even where machinery was not an issue. The Crispins engaged in no machine breaking, and the strikes that did take place against working with machinery were aimed not at the machine itself, but at the use of "green hands" to operate the machines in place of skilled workers. Thus control of the supply of labor became the instrument by which innovation was to be met.

The introduction of the McKay pegging machine in the 1860's sharply altered the shoemaking industry. The machine greatly simplified produc-tion of shoes, allowing for increased division of labor and the replacement of skilled hand craftsmen with unskilled workers quickly trained as machine operators. The apprenticeship system became unnecessary because, as Ira Steward put it, "Instead of wasting so many years to make a few men ready for shoemaking, shoemaking is made ready for anyone."[16] Shoemakers reported to the Massachusetts Bureau of Labor Statistics that the new ma-chinery reduced the work force by from 25 to 50 percent, reduced the value of skilled labor, created monotonous work, and made it impracticable for workers to open their own businesses because of the cost of machinery.[17] This was to be a familiar indictment in later decades as the use of machin-ery widened.

The initial reaction of some shoemakers, including members of the Crispins, was to resist the machine, and opponents of the Knights tried to label the organization as one opposed to mechanization and progress.[18] Despite some opposition to the machine from local lodges, most of the mem-bers of the Knights, as well as the leadership, rejected opposition in the firmest terms, and instead sought to check the introduction of "green hands" into the trade.[19] Thus the Crispins hoped "to share themselves in the pecuniary advantages obtained from the full use of machinery."[20]

Control over the supply of labor was justified in the traditional terms, with little attention to the effects of the McKay machine as such. Thus it was argued that the Crispins had been founded "on an assumption that there was at that time an excess of labor over the demand for it."[21] Since 1867, conditions had worsened, and to remedy this the worker had the same right as the capitalist not "to impart his skill, which is his capital—to be used in injurious competition against him."[22] In terms of the threat from the machine, which reduced the importance of that skill, the worker also had a right not to work with the unskilled, who clearly were competitors for employment within the trade.[23] The Crispins thus sought to restrain the entrance of new workers into the trade in order to conserve employment for the existing work force.

Skilled workers might still find a place in the trade, operating the machines, if a new work force were not recruited; and with control over the supply of labor, the Crispins would be able to resist the reductions in wages that employers expected from the use of unskilled labor. Although some craftsmen would have preferred to oppose the machine, the Crispins, and the labor movement generally, recognized that strikes against machinery usually failed. Instead, labor leaders sought to apply the well-established principle that control of the supply of labor placed the trade union in a favorable bargaining position to get the best terms possible in the new conditions of employment created by technological change.

Until 1869, this objective was aided by support from smaller manufacturers in the shoe industry, who also were threatened by mechanization. The introduction of machinery was often disastrous for smaller manufacturers, who lacked the capital to convert to production in a factory. The campaign of the Crispins against green hands promised to keep the wages of machine operators at a much higher level than if the unskilled replaced the existing work force, and this would benefit the smaller employer by reducing the spread between his own wage costs for hand labor and the cost of labor in a factory.

Support from small businessmen helped the Crispins gain in influence during 1868 and 1869, but it waned after that date. The change of attitude resulted from the Crispins' insistence on concessions from the smaller manufacturers as well as opposition to green hands. Many smaller manufacturers found that the advantages gained by support of the Crispins' position on unskilled labor and mechanization were offset by concessions that they had to make to their own skilled employees.[24]

Cooperation between employers and unions to combat new developments took place in other industries as well. As we have observed, some coal mine operators briefly supported the efforts of the miners to control production in the late 1860's. Furniture manufacturers in Chicago called for eight hours to reduce production during the depression of the 1870's.[25] In the late nineteenth century, the Master Plumbers of Brooklyn shared their journeymen's interest in restricting apprenticeship. The employers feared that an oversupply of journeymen would eventually lead to unemployment for some of the skilled workers, who might then become low-priced contractors, increasing competition in the trade. Similarly, employers in the horseshoeing trade sought to enforce apprenticeship regulations.[26] In both cases, the employers operated small concerns, and they feared that a disruption of traditional labor patterns might work to their disadvantage. Despite such examples, cooperation between employers and trade unions to preserve existing methods of work was not a major feature of the period. Although some employers supported certain objectives of trade unions, they also feared the power that a restricted supply of labor and a strong labor organization would give to their own workers.

Although trade unions claimed that they would limit the supply of labor only as far as demand required, employers often faced the situation of the pottery manufacturers in East Liverpool, Ohio, who were faced by a decision of the potters' union not to increase the supply of apprentices even though an expansion of business required it. The union insisted on maintaining the smaller number of men because experience indicated that apprentices would be used instead of journeymen should the volume of business later recede.[27] In the highly organized window glass industry, however, President James Campbell set aside his own union's rules limiting the number of apprentices in order to meet the need for more workers.[28] The trade was well organized, and collective bargaining with employers was well established, which allowed Campbell to create a true balance between supply and demand. Few trade unions found themselves in this enviable position, and thus the effort of employers to expand the supply of labor was usually met by a union's attempt to restrict it, with little real attention paid to the actual needs of the trade. Thus the mutuality of interest produced by a common fear of new conditions was usually too fragile to resist for long the traditional antagonisms of the wage system.

In the last quarter of the nineteenth century, the issue of the supply of labor in an industry continued to find expression through the apprenticeship question, even though it was recognized that apprenticeship was "nearly if not quite obsolete."[29] If a union could establish regulations on the number of apprentices permitted in a trade, it would be one method of reducing the impact of industrialization. A smaller flow of new workers into the trade improved the chances that those already in the labor force could secure employment at favorable wages despite technological innovation.

The actions of specific unions illustrate the importance and function of apprenticeship rules. Technological change greatly affected the carpenter as building components, such as doors and sash, once made at the construction site, were manufactured in factories. The carpenter then installed the components, but this made it possible for many "hatchet and saw" men to enter the trade, since the skill needed was greatly reduced.[30] The Maryland Bureau of Labor Statistics estimated in 1896 that about half of the carpenters in the state could be designated as relatively unskilled men.[31] The United Brotherhood of Carpenters and Joiners viewed this development with alarm, and apprenticeship limitations were a prominent feature of the work rules established by its locals. The national union stressed immigration restrictions as another means of slowing the influx of men into the trade.

Painters complained of the ease of entry into their trade even before the appearance of the spray gun at the end of the century.[32] The Brotherhood of Painters and Decorators set a definite apprenticeship period in 1888, but left the proportion of apprentices to journeymen for locals to decide.[33] As with the carpenters, this reflected the local nature of much of the building construction industry.

At the end of the nineteenth century, iron molders struck to prevent the expanded use of semiskilled workers from destroying apprenticeship rules and thus weakening the ability of all workmen in the trade to bargain effectively with the employers.[34] Machinists resisted the increased use of helpers, who were able to learn the simplified processes that developed in this trade, and then threaten the jobs and wages of the existing journeymen.[35]

The tailors not only faced division of labor in their craft, but also found employment dispersed into innumerable sweatshops and home workshops rather than concentrated in factories. While anxious to control entry into the trade by restricting apprenticeship, the major effort had to be directed at concentrating production in factories. Until this was done, the apprenticeship regulations that were passed periodically had little effect.[36]

The division of production in the hat trade allowed boys to be substituted for men in many processes. The John Stetson Hat Manufacturing Company had begun to use boys in place of men in its operations as early as 1876. In 1885, Assemblies 3535 and 3560 of the Knights of Labor called on American workers to boycott Stetson hats because the number of boys trained by the company had not only driven many men from its employ, but now threatened to flood the entire trade with "boy labor."[37] Apprenticeship regulations were one way to meet these threats to the established work force. Where effective, they allowed the trade union to control, by regulation, what had once been limited in large part by the requirements of skill—the ease of entry into a trade.

The discussion of apprenticeship also raised important questions about the means that the workers, or their trade unions, might use in the defense of their vital interests. In 1874, a former coal miner wrote that his fellow workmen in the pit "feel they have an actual property right to their place," and when the changes in the labor force between 1869 and 1873 displaced some older miners, they had the "same emotions that an ordinary person would if robbed of his home."[38] In 1883, M. D. Connolly, a printer from Covington, Kentucky, argued that he regarded his skill "in a certain sense as a vested property right," to be exercised for one's own good. "Hurtful competition" from those outside the trade was dangerous to the printers' interests, and it had to be prevented by strict limits on the number of apprentices.[39] As a miller complained in 1880, without regulation of entry into the trade, less qualified workers would ask only half the wages that those then at work received. "The proprietor does not hire him but he uses him to force the good workman to submit to a reduction of wages." Therefore, men should be examined and certified as qualified before entering the trade.[40]

The union of glass bottle blowers also claimed the right to close entry into a trade in order to protect the interests of those already at work. In 1886, the union demanded the cessation of all new apprenticeships because only two-thirds of the journeymen were able to find work. The surplus was blamed on the employers' past policy of expanding the labor supply as a means of undercutting the bargaining power of the skilled workers.[41] The leading trade unionist, Frank Foster, likened the position of the worker to other commodities in the marketplace. Thus if Standard Oil might fix the price for oil, by controlling the supply, workers might similarly regulate the supply in order to achieve a favorable price for their labor.[42]

Delegate L. W. Tilden put the issue even more bluntly before the convention of the plumbers' union in 1897. "A labor organization is a monopoly from start to finish, and we are entitled to have it just as much as the coal and sugar trust is entitled to monopoly." Such a statement was rarely made by labor leaders, for as delegate John S. Kelly remarked, however serious the evil of the helper or apprentice, "the public has got to be taken care of and considered."[43] Instead, Kelly believed emphasis should be placed on organized labor's traditional objective: limit supply to demand in the trade in order to avoid unemployment and wage reductions.[44]

Opponents of organized labor insisted that apprenticeship rules were proof of the basically antisocial purpose of trade unionism. However, defenders of organization among workers pointed out that apprenticeship practices and rules reflected the realities of the existing economic system. Thus Commissioner Charles Peck of the New York Bureau of Labor Statistics acknowledged that apprenticeship regulations were "actuated by selfishness," but he likened them to the limits placed on entry into the learned professions.[45]

L. G. Powers, Commissioner of the Minnesota Bureau of Labor Statistics, argued that the greatest industrial problem of the age was how to secure the immense benefits of technological innovation for the community without injuring any portion of that community. He recognized that such a reconciliation had not taken place in the English Industrial Revolution, nor was it automatic anywhere. The labor movement was a response to such industrial change. Unions "exist to aid their members, first, to avoid the hardships and correct the evils incident to modern industrial changes, and second, to secure for those members a just share in the benefits and gains resulting therefrom." One method of accomplishing these objectives was apprenticeship regulations.

Powers stressed that the apprenticeship, in an industrial situation, was basically a device for introducing cheap child labor, and not a method of training workers in a skilled craft. Thus employers were as little interested in training "apprentices" as labor unions were favorable to the unlimited entry of workers, especially children, into the trade. Unions attempted to restrict child labor so as to "guarantee that industrial changes due to the use of such labor will work them the minimum of harm." The end result was that "the fight is made on both sides without any primary regard for the interests of the child or society as a whole. Each party to the conflict seeks to advance his own interests and those of the class to which he belongs." It was the state which had to adjudicate such conflicts so that helpless parties, such as children, were protected while the overall interests of the society were preserved.[46]

Powers noted with approval a decision by Judge Tuley of Chicago in 1887. In a controversy over apprenticeship regulations between the master masons and their workers, the judge refused to view the issue, as conservatives did, in terms of the absolute right of a child to learn a trade. Instead he argued that also at issue was whether "a craft will teach a boy a trade to its own destruction." Determining when, and if, that was the case in any instance should be done by a "joint arbitration committee." Beyond this, Powers believed that the state should legislate in the areas of apprenticeship regulations and child labor, until the interests of the child were fully protected and "regardless of the possible profits of the employers, the possible gain or support of parents from the labor of the child or the possible claim of any body of workers."[47]

Powers's view of trade unionism and its relationship to government was an early statement of the basic position developed by the Social Progressives during the following quarter century.[48] Like Powers, they stressed the legitimacy of trade unions, which were necessary for the protection of workers. Yet they also insisted that self-interest, however justified, could not be presumed to be identical to the national interest, which ultimately had to be served by an impartial agency—the state. Many trade union leaders, including Samuel Gompers, were no more ready to accept domination by the state than they were ready to acknowledge control by the employer.

In large measure, organized labor's continued use of the old apprenticeship device was made necessary by the lack of satisfactory alternatives. Control of the labor supply had been a basic principle of trade unionism, and industrialization made such control all the more necessary at the same time that it threatened the skill which had been the traditional basis for limiting the number of workers. The state's power to curb the negative effects of industrialization was limited by the generally accepted principle that the working conditions of adult men were not subject to legislation. Moreover, by the end of the century, many labor leaders distrusted actions by governments over which they had little influence. Apprenticeship was one well-understood method for controlling the supply of labor, and it was used for this purpose even when it no longer was a viable institution for training new workers. Thus trade unions turned to familiar devices, which they used as long as such practices promised to offer some degree of protection against the insecurity of industrial change.

## CHAPTER EIGHT

# FEMALE AND CHILD LABOR AND INDUSTRIAL CHANGE

Closely related to the apprenticeship issue was the response of American labor to the increasing numbers of female and child workers.[1] Obviously, the attempt to establish limits on the number of apprentices was one means of controlling the entry of children and young adults into specific trades, but this approach could be supplemented by legislation which would regulate the flow of children into the work force as a whole. The labor organizations of the 1830's and 1840's had sought such laws, and efforts continued in the late 1860's and 1870's.[2] Organized labor continued to press for child labor legislation during the remainder of the century, and a number of state laws were enacted. However, they were weakly enforced. These laws had three basic features, although the details varied from state to state and from year to year. First, child labor, under a specified age, was totally prohibited; second, children above the specified age were permitted to work, but their hours, or the trades in which they could labor, were regulated; third, some minimal amount of education was prescribed.

These regulations received considerable support outside the labor movement. Children could not protect themselves, and thus the state had to insure that their health would not be adversely affected by labor at too early an age, or for too many hours, or in dangerous occupations. The community's interests also were not served when child labor created a body of future citizens who were illiterate or barely able to read and write.

For American workers, these evils merely supplemented the basic one: the destructive effect of children upon the conditions of labor for adult men. This was true in the period from 1850 through the 1870's as the division of labor increased the opportunities for children to work effectively in competition with adults. The rapid mechanization during the last two decades of the century made it possible for children to hold a much wider range of jobs. After some brief training on a machine, a young worker would often operate it as efficiently as an adult, yet the wages paid children were considerably lower. Child labor legislation would lessen the opportunity for children to enter the labor market, and, together with attendance laws, increase the time spent in school.[3] Legislation could also reduce the profitability of child labor, in comparison to adult, by placing limitations on the hours, or type of work, that might be performed. Labor leaders often cooperated with reformers on the child labor issue, and the humanitarian concerns usually received primary attention in campaigns for public support and legislative action. Yet it is quite apparent that American trade unions were most concerned with the destructive effects of child labor upon adult workers. Organized labor sought to reduce work by children as part of its broader effort to control the supply of workers as fully as possible.

American trade unions regarded labor by women in much the same way. Even where skill had not been challenged, trade unions were hostile to the introduction of female workers. The National Typographical Union debated the issue in 1854. The Detroit local called for opposition to female compositors because they were used by employers "to set aside fair wages and compensation"—a reference to the already well-established practice of paying women less even if they did the same work as men. A proposal was made to admit all women, who had served a regular apprenticeship, to the union, but it was defeated in favor of a resolution "That this Union will not encourage, by its acts, the employment of females as compositors."[4]

William Sylvis argued that the "defenseless condition" of women led them to work for lower wages than men, thus placing pressure on men to accept less. "If they received the same wages that men do for similar work, this objection would in great measure disappear."[5] Yet the labor movement made little effort to achieve such equality since, in reality, the basic objective was not to win equal wages for female workers, but to prevent women from competing with men for jobs.

Accordingly, Jonathan Fincher opposed the introduction of women into trades that contained male workers because they would not only lower wages but replace men. Fincher argued that women were used only when they could be paid lower wages, and thus the equality of wages theme was unreal.[6] He also attacked the unsexing of women, which resulted from female workers leaving their proper place in the home. Since women were already at work in significant numbers in some trades, he called upon them to restrict their labor to these trades and to organize their own labor unions.[7] Despite the statements by some labor leaders that seemed to accept women, if they were paid the same wages as men, the actions of trade unions revealed that Fincher's position was the dominant one.[8]

Mechanization was to intensify the problem of female workers just as it made child labor a more serious threat. Women could operate many machines, thus opening trades that formerly had been worked only by men. The labor movement responded by extending its old fear of female skilled workers to the unskilled machine tenders who might happen to be women. Thus legislation designed to limit female labor won wide support among the trade unions. Again reformers were often leaders in this area. They usually stressed the negative effects of industrial work upon the supposedly frail constitution of a woman, and the threat to her moral fiber from conditions in the factory. Perhaps most important was the fear that female labor would destroy the family structure and thus weaken the basic fabric of society. Labor leaders certainly accepted such arguments, but clearly their basic concern was the effect of women on the industrial system and not the impact of the factory upon women.

Despite the claims from some that women and children did not compete with men for jobs, it became almost axiomatic in the labor movement that women and children displaced male workers and lowered wages. In 1878, the labor editor George Gunton viewed the introduction of women and children into industry in New England as a systematic policy to supplant men. Not only were women less expensive than men, but they were also more difficult to organize. Gunton called for laws against child labor and a heightened effort to organize women into trade unions.[9]

Some labor leaders stressed the highly competitive industrial system that led employers to seek out the cheapest labor possible in order to achieve an advantage. Uriah Stephens, the founder of the Knights of Labor, pointed out in 1879 that "Perfected machinery persistently seeks cheap labor and is supplied mainly by women and children. Adult male labor is thereby crowded out of employ, and swells the ranks of the unemployed or at best the underpaid." Stephens thought equal pay for equal work would cut the attractiveness of female labor, and it was necessary to prohibit child labor entirely. "It becomes an alarming symptom of demoralization and degradation of the race when childhood is defrauded by greed out of education and healthy development, and the motherhood of the next generation made machine-slaves."[10]

It was also recognized in the 1870's that unemployment in any trade had a snowballing effect, since the men forced out might well seek jobs in other industries. As skilled hand labor declined, generally, the barriers of craft that had protected a trade declined. Thus every trade where skill had been diluted eventually felt the effects of the displacement of men by women and children.[11]

Before the congressional committee that had been established to study the depression of the 1870's, Charles Litchman, a leader of the Knights of Labor, testified about the effects of machinery in shoemaking. Men had been replaced by women and children, creating adult tramps in the present generation and laying the seeds for the degradation of these self-same children in the future. Litchman would keep children in school, through compulsory education laws, even though he recognized that "selfishness on the part of the poor will sometimes compel them to violate the law."[12] Poverty was so basic a reason for child labor that Commissioner Charles Peck of the New York Bureau of Labor Statistics regarded it as more important than mechanization as the root cause for the increasing number of children found in industrial labor.[13] Child labor laws passed in the nineteenth century were generally ineffective, not only because employers sought to evade them, but because parents did so as well. The state failed to provide the personnel to enforce these laws in the face of widespread evasion. Eugene Debs agreed with Commissioner Peck's conclusion, and he pointed out that while child labor laws might cut down on one form of cheap labor, they did nothing to alleviate the basic reason for it. The end of child labor—without a general improvement in the wage level of the father—could lead him to send his wife to work instead.[14]

Poverty may have provided the need for child labor, but the machine provided the opportunity. Thus the president of a local cigar makers' union linked child labor directly to the machine, which allowed children to replace men in the work force. For him, the answer was not child labor laws, which regulated only the effects of mechanization, but cooperative cigar shops which would use machinery to serve the interests of workers rather than the profits of employers. Such shops would refuse to employ boys in place of men. "Then we could let the machine do the work and we would receive the benefit of the same, but as it stands now it is a curse to us all."[15]

Reformers also stressed that female and child labor not only displaced many adult men, but also had a disastrous effect upon the families involved, the general community, and even the employer. The popularity of a consumption theory, as an explanation for the recurrent depressions that accompanied industrialization, led to the conclusion that cheap labor was actually contrary to the interests of the employer. Child labor cut the power to consume by substituting the lesser earnings of a child for the wages of the father. At the same time, production increased through mechanization. This created a dangerous consumption gap that threatened the economic health of the entire community. The end of child labor would increase the wage bill for employers, but would also increase the market. The greater efficiency of an adult meant increased productivity per man-hour, so that the employer's profits ultimately would be greater through the use of more highly paid adults than through the use of cheap child labor.[16] No single employer could make such a choice in the face of short-run competitive demands, and so law had to set the rules so that no concern could secure a competitive advantage.

The opponents of child labor were unanimous in stressing its depressive effect on the wages of adults. Those who accepted some variety of a standard of wants theory as a basis for wages also argued that child labor could do little to raise the family's income. The reform economist William F. Willoughby stressed the "standard of comfort" in his discussion of child labor. Like Ira Steward, Willoughby denied Ricardo's iron law of wages or any variety of a wage fund theory; instead he insisted that wages were paid upon the customary level of comfort expected by workers. This would naturally vary from nation to nation and from time to time. Child labor did not increase the existing level of wages because the family's standard of comfort did not rise. Thus it would accept collectively what would have been earned by the father alone. In fact, child labor would ultimately lower a family's income, no matter how many persons worked, because labor during a child's formative years prevented education and in every other way limited the child's horizons and experience. Accordingly, the young worker's standard of comfort declined. This situation also created an intergenerational downward pressure on wages because the deprived child expected less as an adult.[17] The end of child labor would lead to an increase in the wages of men to a level approximately that of the former earnings of father and child.[18]

Child labor also created a vicious cycle: the labor of women and children lowered the wages of men, which made it necessary for workers to send their families into the factories, thus lowering wages for still other men and extending the evil effects of the system.[19] In 1875, the Massachusetts Bureau of Labor Statistics examined the contribution of women and children to the total income of families. Although the proportion added by wives was small, children supplied one-quarter to one-third of the family's income, with those under fifteen supplying one-eighth to one-sixth of the total. Thus Commissioner Carroll Wright had to conclude "that without children's assistance, other things remaining equal, the majority of families would be in poverty or debt."[20] The vicious cycle could be broken only by restricting child labor, which would allow adult wages to rise, thus freeing the father from a dependence on the labor of his children and allowing him to educate them so that they would have a greater measure of economic opportunity in their adult years.

The dependence of a father upon the earnings of his wife and children could have an impact beyond the issues of employment and wage rates. Neil Smelser has pointed out that the English artisan accepted the value placed by society on economic independence and the associated virtues of frugality, self-reliance, and hard work. As breadwinner for the family, he was responsible for maintaining its economic independence. In the process he maintained control over the family and his status in the community. However, as the British cotton industry moved to the use of machinery in factories, female and child labor was increasingly utilized. Although women and children had often contributed to the family's income, their wages had been a supplement to the income of the father. Mechanization often reversed the situation, and men could be unemployed while their wives and children found employment tending the new machines. Even if the father continued to work, the proportion of the family's income produced by the father could decline significantly. In such a family, the father's role as breadwinner was destroyed, with a consequent loss of status in the community.[21]

Smelser's conclusions are suggestive for understanding the situation in the United States. Nineteenth-century America not only placed a positive value on the economic independence associated with the role of the father as breadwinner, but tied this closely to the mobility ethic. Economic independence was highly valued in itself, and the theme of an increasing dependency was an important feature of the general reaction to industrialization.[22] Moreover, the desirable social model was a family supported by the father.[23]

On the basis of his study of a Chicago neighborhood, Richard Sennett has suggested that the "wives of lower white collar and blue collar workers" worked less than might have been expected on the basis of the financial needs of the family. He concludes this was a result of the challenge posed to the primacy of the husband as head of the family, since wives might earn as much as their husbands.[24] However, among unskilled workers generally, female and child labor was quite common, and critics of immigration in the late nineteenth century stressed labor by women and children as one of the most undesirable characteristics of the newcomers.

Of course, women and children had worked before the advent of the mechanized factory. In fact, proponents of unrestrained mechanization claimed that fewer children were at work in an industrial society than in a more primitive economy. The increased wealth of the nation, including the higher wages earned by workers, allowed more children to spend more years in school.[25] However, in an economy of small workshops and skilled workers, a child was often apprenticed in his father's trade, and many times even in his father's shop. Such an arrangement maximized the control of the father over the child. Labor by children in an industrial system usually meant that the child found employment in a totally different situation from his father. In addition, the father might suffer the indignity of being unable to find work while his wife and children were employed. By reducing the economic primacy of the father, female and child labor also created a gulf between such families and the social ideal. At the same time, the ability of a worker to move into the entrepreneurial ranks had diminished sharply, further reducing the status of the worker, which had always been based upon possible access to the middle class rather than upon any inherent respect for workers as such. Thus the skilled worker not only faced the economic threat of child and female labor, but profound social effects as well.

Those who called for an end to child labor recognized that it might create an immediate hardship for families that were dependent upon their children's earnings. However, William Willoughby argued that any social change injured some portion of the community. Craftsmen suffered from innovations that reduced the need for their skill, yet "improved machinery is not forbidden on that account."[26] Although Willoughby was using this argument to try to destroy the image of the child supporting the impoverished widow—a device long used by conservatives as part of their claim that child labor was essential to the poor and must be maintained on their behalf—it still left the worker with no shield against the immediate economic effects of ending child labor. To overcome this problem, some supporters of reform demanded that the state provide the cushion necessary to abolish the evil of child labor without causing still further distress for those who already suffered from this social curse.

In 1875, Carroll Wright called for the prohibition of all child labor below the age of fifteen, and a system of free common schools that would integrate children from all social classes. This last feature once again shows the influence of Steward's contention that the standard of living was ultimately based upon wants and expectations, which would certainly rise if the children of the poor could observe the habits of the middle class. To aid parents who suffered from the loss of a child's earnings, Wright would have provided state financial aid.[27] Wright also suggested a minimum wage for adults to help meet the broader problem of poverty, one result of which was child labor.[28] In 1883, William Weihe, a leader of the Amalgamated Association of Iron and Steel Workers, made much the same proposal concerning financial aid to poor families forced to give up the proceeds of child labor. He felt that the end of child labor would allow more men to provide for their families, but if children faced starvation, "I would have the Government clothe them and provide for them where it was necessary."[29]

Clearly action by the state was one type of response to the problems raised by industrialization, but trade unions in the 1880's and 1890's did not view extensive action by governments as the answer to industrial change. In those areas where experience indicated that employers would negotiate with strong trade unions, the labor movement asked little help from the state. Of course, such a policy left the unorganized workers with little protection, and limited the effective area of response to the most strongly organized trades, and eventually, in most cases, to the strongest locals within the union.

Yet in some areas state action seemed desirable because trade unions could not act effectively, or public support for specific legislation promised not only passage of a law but some hope for enforcement. Regulation of work by women and children and restriction of immigration were important areas in which trade unions expected government action. English precedent sanctioned the use of law to regulate the hours of women and to prohibit or limit the hours of work for children. Whatever the constitutional issues concerning the regulation of the labor of adult men, the state seemed to have the right to protect women and children from the effects of the factory system.[30] For children, the loss of educational opportunity and the real dangers to health from long hours in the factory were sufficient reason for the state to intervene. For women, the negative effects of factory labor upon their supposedly frail natures also permitted action by the government. Thus labor leaders might view the issue as one of competition for employment under the aegis of increased mechanization, but support for legislation regulating the labor of women and children did not have to be placed before the public and the legislatures on this basis. Instead the stress could legitimately be upon the ill effects of labor upon women or children rather than the evil effects of their competition upon adult men.[31] This allowed for a humanitarian appeal that could rally support from those who would be far less interested in the problems of the labor movement.

As we have observed, in 1875 Commissioner Carroll Wright of the Massachusetts Bureau of Labor Statistics proposed that no child under fifteen years of age be employed. He also suggested that the hours of labor for women be adjusted to fit their special physical needs. Wright believed that overly long hours, combined with the excessive speed and concentration demanded in many industries, sapped the strength of many women. The list of occupations that posed such health hazards was wide, and it included most of those in which women competed actively with men or threatened to do so. Thus factory labor, typesetting, telegraphy, sewing machine operation, and the manufacture of tobacco all supposedly produced health problems for women.

Wright made it clear that "If such an expression would not be considered as bordering on the insane, we should say at once, that married women ought not to be tolerated in mills at all. Vital science will one day demand their exclusion; but *we* certainly can recommend the regulation of their work." Such regulation included safety and health codes for factories, limitations on the hours of labor, and enforcement by inspectors.[32] The proposal thus contained all the major features of the laws controlling labor by women that were passed in the Progressive period of the early twentieth century.

Legislation that protected the health of women tended to make their labor more expensive, and this could reduce the competitive advantage that women often had in the labor market. The smaller the gap between the cost per hour for employing women and the customary wages for men, the less likely the displacement of adult men or severe pressure for wage reductions. Another device to achieve the same end was simply to insist on equal pay for equal work.[33] However, this approach had the disadvantage of seeming to welcome women into the labor force, whereas legislation on hours and conditions of labor for female workers clearly carried a measure of disapproval. Regulation might lead ultimately to prohibition of at least certain categories of female labor as was proposed in the case of children. Also equal pay lacked the humanitarian appeal of legislation designed to protect the health and welfare of women or children.

Factory legislation designed to conserve the health of women could also have a very real effect on the working conditions and hours of men. Since women and men often worked in interrelated jobs, a limit on the hours of women would by necessity fix the hours for men. Health codes for women in factories often required facilities that would also improve the working conditions of men. Thus legislation for women could indirectly provide for men the very conditions that would be frowned upon if proposed as a proper function of the state.

Considering the importance of legislation on female and child labor, and the seeming availability of allies outside the labor movement, it is not surprising that labor organizations played an active role in trying to secure the necessary laws. State federations of labor lobbied for such legislation, and the American Federation of Labor tried to use its influence in the same direction. The question of means soon came to the fore. It became evident that action by the states would be extremely slow and spotty, and the resulting variety of laws tended to limit the gains to the pace set by the least progressive states. Employers were more prepared to accept legislation if they believed that no real competitive disadvantage would occur. Similar laws in the leading manufacturing states would prevent one group of firms from gaining the economic advantages of employing women and children more freely and would also allow increased costs to be more easily passed on to the consumer.

The advantages of action by Congress were thus obvious, but the states were the traditional source of legislation concerning the labor of women and children. To partially overcome this problem, the American Federation of Labor, in 1890, proposed a constitutional amendment that would allow Congress to prohibit the employment of children under fourteen in mines, factories, or stores.[34] It was not until 1916 that such action was taken by Congress and then by statute without a constitutional amendment. This law, and a subsequent one, were declared unconstitutional, which led to a serious effort to pass a constitutional amendment authorizing Congress to legislate concerning child labor. Although the amendment passed Congress, it never secured approval in the requisite number of states, and federal regulation of child labor did not occur until the New Deal.

The difficulties in securing a constitutional amendment during the 1890's led the American Federation of Labor to return to an emphasis on action in the states. In 1894 the convention of the A. F. L. adopted a resolution calling for "a uniform limit to the hours of labor for women and children in all manufacturing establishments" to be secured by action in the state legislatures. In addition, compulsory education laws should be passed in all the states.[35] The A. F. L. did not join the woodworkers' union, which called for a penalty of five to twenty years in prison for violation of the child labor laws,[36] but it made a determined effort to secure effective regulation of the labor of women and children.

That the accomplishments were slim was not the result of a failure in concept or will, nor was it a function of the political isolation of the labor movement. Allies among reformers, and even within the ranks of conservatives, were available on this issue. The crucial factor was simply the superior political influence of business interests as compared to the alliance favoring reform. Although organized labor continued to provide strong support for legislation in this area, it was the organization of the reformers into effective political lobbies during the Progressive period that was most important in securing and enforcing meaningful child labor legislation.

Although child and female labor were important issues in themselves, basically the labor movement regarded them as another element in the increased competition among workers, which, in turn, resulted from the effects of industrial change upon the traditional protection provided by skill. Thus there was no basic difference in purpose between labor's campaigns for the regulation of female and child labor and the calls for restrictions on immigration or apprenticeship. Cheap labor was a threat, whatever the source. Those outside the labor movement often concentrated on the particular reform, not the basic issues of unemployment and lower wages. Organized labor's demand for child labor legislation gained a large measure of public acceptance, but the call for apprenticeship regulations was attacked as self-serving. The hours of women could be regulated and reduced, but not the hours of men, even though the labor movement regarded the latter as more basic in meeting the threat of industrial change.

Some observers recognized that workers viewed mechanization as the ultimate threat to their interests. It could displace skill and throw workers into a seemingly never-ending scramble for employment. Commissioner W. H. Hilkene of the Missouri Bureau of Labor Statistics wrote in 1880 that a man displaced by a child, in the prime of his life and after devoting years to acquiring the knowledge necessary to his occupation, would regard machinery as a curse rather than a blessing.[37] As Commissioner Charles Peck put it: "This lack of employment, this unsteady and irregular work, is, I repeat, at the bottom of it all. Is it to be wondered at that the unions have felt constrained to throw up barriers and place restrictions around their trades in attempts to prevent those out of work from competing with those at work."[38] Such curbs included restrictions on female and child labor and control of apprentices, but equally important were the demands for barriers against immigration and the campaign for a reduction of hours.

## CHAPTER NINE

# EIGHT HOURS AND INDUSTRIAL CHANGE

Of all the measures designed to meet the challenge of industrialization, shorter hours had the widest and most sustained support among trade union leaders. Whatever the value of control of apprentices and the regulation of female and child labor, however salutary the restriction of immigration, a reduction in hours was basic in any effort to cushion the impact of industrial change. The eight-hour work day was one of the labor movement's major demands in the late 1850's and 1860's, even though mechanization was less the issue than the division of labor, the breakdown of skill, and the resulting increase in the number of workers in a trade. Before the Civil War, the constitution of the Mule Spinners included among its objects "ways and means to employ surplus members of our Association," and eight hours was recommended as the "best means to improve the trade."[1] The Machinists and Blacksmiths called for eight hours at the convention of 1859, and the Iron Molders' convention in 1861 also demanded shorter hours.[2] Following the Civil War, the eight-hour day became one of the major demands of the labor movement.[3] An eight hour campaign was the major undertaking of the young American Federation of Labor in the 1880's.[4] Samuel Gompers maintained, in his report to the Federation's convention in 1887, that so long as there was a single worker unemployed, the hours of labor were too long.[5] Commissioner Thomas Dowling of the New York Bureau of Labor Statistics noted in 1894 that "throughout all the trade organizations of the State there is a manifest unity of action to reduce the hours of labor to keep pace with the productivity of machines."[6]

In good measure, the trade unions organized in the 1850's and 1860's were a response to the threat that changing trade conditions would undermine the role of skill as a regulator of the supply of labor. The skilled craftsman feared innovations that opened the trade to a large number of workers and increased the supply of labor without any guarantee that the demand would increase proportionately. In addition, the increased number of workers remained in the trade when business turned slow, either seasonally or cyclically, and thus the bargaining power of workers was reduced by the increased competition among them for the available work. The unemployed not only undercut the efforts of workmen to secure the highest wages possible, but their presence made available a pool of workers whom employers could induce to become strike breakers. In the capitalist economy of nineteenth-century America, there was no means of balancing the number of workers in the trade with the reasonable expectations for labor: in fact, it was clearly in the interests of the employer to create as large a supply of available workers as possible.

Machinery was an even greater danger because it speeded up the destruction of skilled labor as the method of production, and, in so doing, removed the major weapon used by craftsmen to keep the supply and demand relationship favorable to their interests. Thus machinery was not treated as a new enemy, but as the latest and most dangerous element in the continuous process of division of work and dilution of skill. Labor did not develop new weapons to meet the challenge of the machine because it did not perceive mechanization as a fundamentally new process. Instead, it was the ultimate weapon of the employer in undercutting skill, flooding the labor market with semiskilled and unskilled laborers, and thus creating a permanent surplus of workers. To labor leaders, the call for shorter hours was primarily designed to reemploy idle workers and thus absorb any oversupply of labor. This would restore the bargaining power of workers and their unions.

The key to the effectiveness of shorter hours was also its most controversial feature. Trade union leaders envisaged that a reduction of hours from ten to eight would mean that a 20 percent larger labor force would be required to produce the existing quantity of goods. Moreover, workers would receive the same wages for eight as for ten hours of labor.[7] Such a program struck at the very heart of the economic theory accepted by conservatives and most reformers, as well as the practice of businessmen, which based progress upon the increasing productivity of workers. Machinery allowed each worker to produce more per hour of labor, which reduced the cost of production, and ultimately the price of the object, while increasing the supply enormously. The combination of an enlarged supply of goods, at greatly reduced prices, created the opportunity for the masses of the population to consume products once available only to the few. A mass production industrial economy thus depended upon increased productivity per worker, and any proposal to reduce hours without a corresponding increase in output per worker was considered a threat to the economic progress of the American people.

Labor leaders understood this argument, but rejected it despite the opposition raised by their position. In his testimony before the Senate Committee on Education and Labor in 1883, Robert Layton, Grand Secretary of the Knights of Labor, made it clear that "A man working eight hours a day will not do as much as a man working ten hours, and therefore if you shorten the hours of labor you give the surplus labor an opportunity for employment." At the same set of hearings, Adolph Strasser, President of the Cigar Makers' International Union, was asked whether unions believed in suspending work periodically in order to lessen production and prevent industrial panics. He responded that labor sought to increase the consumption not lessen the supply, but if machinery increased production too rapidly, a reduction in hours would decrease total output. However, Strasser accepted Ira Steward's belief that increased production could be expected in the future once eight hours generated new wants, through increased leisure, and thus expanded the demand. Production could then grow without the threat of overproduction.[8]

In the eight hour primer that he prepared for the American Federation of Labor in 1888, George McNeill stated unequivocally that "The day will never come when one can do as much in eight hours as he might the day before or after in ten hours' labor." However, this was no reason to refuse the demand of workers for the shorter day.[9]

The reason for the labor movement's position is clear. Trade union leaders regarded shorter hours primarily as a device to reduce unemployment. They believed any increase in the productivity of workers meant that fewer men would be needed to provide the existing production. Opponents of this view argued that the amount of work available could be expanded rapidly through technological gains that lowered prices and generated new demands. However, trade union leaders rejected this conclusion because they regarded unrestrained innovation and mechanization as the very factors that displaced skilled workers, created continuing and increasing unemployment that reduced total demand, and thus ultimately were responsible for overproduction and depression. Therefore, society should have no concern over the fact that shorter hours would not be linked to increases in the existing level of output. Society first had to overcome the existing negative results of uncontrolled industrial change by restoring a balance between work available and workers available. For this purpose, a reduction in hours without an increase in productivity was an essential mechanism.

In the future, a reduction in hours would also be the basis for a healthy increase in output. The shorter work day would put more persons to work and increase consumption. This would absorb the existing overproduction in the first instance and ultimately allow for an increase in output.

Organized labor's position on the issue of productivity and hours was clearly expressed in 1878 by H. J. Walls, the first Commissioner of Labor Statistics for Ohio and a former Secretary of the Iron Molders' International Union. Walls argued that it was fallacious to expect, as many did, that with more efficient machinery, workers could produce as much in eight hours as they formerly did in ten. However, "even if true [it] would destroy the principal claim why the hours of labor should be reduced, i.e., that through the increased power of machinery there is not work for all. Machinery, it is claimed, causes the increased production and the decrease of workers. If the worker with machinery produces as much in eight hours as he will in ten, it must follow that the number of workers will not be increased because of the reduction of hours."[10] A bricklayer put the issue more simply over two decades later: "What is the good of cutting down the hours if we are to lay as many bricks in the shorter hours as we did in the longer hours? That will not put any more men to work."[11]

Basically, the position of the trade unions was an expression of their insistence on a fair economy rather than an efficient one. Skilled workmen were victims of unrestrained industrial growth that destroyed an older system of production before it could insure the proper operation of the new one. Organized labor rejected economic theory that based all improvements in the standard of living upon efficiency and unchecked production, regardless of the effect of such developments upon established work patterns and the existing labor force. Instead labor leaders stressed the importance of raising consumption, rather than lowering the cost of production, a position that could justify higher wages and full employment as necessary to the general economic health of the nation as well as the specific interests of the worker.

Despite organized labor's defense of shorter hours, without a reduction in wages and without an increase in output per man-hour, opponents continued to stress the linkage of shorter hours and higher productivity. Employers in mechanized industries rejected a reduction in hours precisely because there was so little room for increasing productivity during the shorter work day. The machinery was already run at the maximum effective speed, and any increase to compensate for shorter hours would lead to an increase in breakdowns or excessive fatigue for workers, with a resultant loss in efficiency.[12] Opponents of a shorter work day generalized this argument to cover all industries. Without an increase in productivity or a reduction in wages, a larger work force would be producing, at best, the same amount of goods in eight hours as in ten, but at a higher total wage cost.[13] This situation would inevitably generate a price rise, which would be to the disadvantage of farmers and other consumers.[14]

An alternative to higher prices was lower profits. Proponents of eight hours within the labor movement were quite prepared that the benefits secured by workers come from the share previously taken by the employer. As an observer pointed out in 1865, profits were already so high that a reduction to even 10 percent would still be a fair return to the employer.[15] Beyond the question of equity, George McNeill contended that a reduction in hours, which inevitably meant an increase in wages, would reduce the percentage of profit per dollar and per worker.[16] However, he denied the contention of conservatives that a drop in profits could lead to the withdrawal of capital, with attendant unemployment.[17] McNeill believed that capital would always accept the percentage it was able to gain and that a reduction in profits would not appreciably modify the amount of capital invested in American industry.[18]

If profits were not reduced—at least initially—it seemed that prices would have to rise. This was a potentially damaging point because it seemed to mean a reduced standard of living for the nation. However, trade unionists believed that even without an increase in productivity, prices would not rise in proportion to the reduction in the workday. In a mechanized factory, the cost of labor was a smaller portion of the total cost than in a small shop. According to one estimate, if the eight-hour day added 20 percent to the cost of labor, prices would increase less than 5 percent.[19] Even more important, to accept the higher price argument as a barrier meant that any improvement in the conditions of workers could be blocked because it raised the cost of production. Improvement in the conditions of labor had a social cost, which had to be met by the nation as a whole. Thus the advantages of shorter hours to workers were mirrored by the wider advantages to society as a whole, and higher prices would be the cost of these benefits.[20]

It also was apparent that low prices need not be a benefit to workingmen, nor an automatic result of industrialization. The labor movement had no interest in low prices if they were the product of rapid mechanization that displaced the existing work force, overproduction that glutted the market and led to lay-offs, or sharp competition among firms in which wages became the major variable in the struggle to cut costs.

Ultimately, organized labor argued that the nation had to break the cycle that tied the standard of living for workers and the economic welfare of the community to unrestrained industrialization. Shorter hours would be a basic step toward such a change in priorities since it would sever the link between progress and productivity. Eight hours would reemploy those who were already the victims of displacement or glut. It might also place a brake on headlong mechanization by insuring that production costs could not be cut without reference to the human results. Eight hours did raise costs and, by so doing, made technological change that displaced workers less profitable. It was to be a continuing process. Should eight hours truly stimulate further mechanization and unemployment, then hours would have to be cut again. Shorter hours was the workers' means of regaining the jobs that industrial change took away. Limitations on output and shorter hours were labor's ultimate response to a society that worshipped technology but ignored many of its consequences.

Many reformers were prepared to accept shorter hours, but not on the basis demanded by the labor movement. They were unconvinced that cutting progress from productivity was feasible in an advanced economy, and they believed the threat to profits, which seemed implicit in labor's eight hour proposals, would be an insurmountable barrier. Thus the key to shorter hours was to find a path that would avoid the problems of economic theory and clashes of interest that were raised by the labor movement's program. Instead, a call for eight hours should rest upon a harmony of interests and the general acceptance of the new, the better, and the more productive.

As early as the 1860's, alternatives to organized labor's concept of eight hours were widely discussed. Ira Steward and his followers argued that productivity per man-hour should increase as hours were reduced. The essence of Steward's theory was that shorter hours engendered an increase in the level of wants, which triggered higher wages. Machinery allowed employers to increase the productivity of their workers, which permitted higher wages and lower prices. At the same time, manufacturers could sell to a larger market and earn good profits. Thus eight hours had to be achieved by increasing productivity through the use of machinery.[21]

One of the supporters of Steward's view in the 1860's was Andrew Cameron, the influential editor of the *Workingman's Advocate*. He accepted the proposition that a reduction from ten to eight hours, without an increase in productivity per man-hour, would raise prices by 20 percent. Such an eight hour movement had to fail. Following Steward, he argued that shorter hours could succeed if accompanied by greater production. The prosperity of workers was rooted in abundance, not scarcity, and the movement for shorter hours had to recognize this fundamental point.[22] Cameron contended further that the normal effect of shorter hours would be increased output. Men would suffer less fatigue in eight hours and thus produce as much as in the longer day.[23]

Many of the commissioners of the state bureaus of labor statistics supported the shorter work day, but once again they refused to do so on the basis demanded by the labor movement. The point was made concisely in 1890 by L. R. Campbell, Deputy Commissioner of Labor for the State of Maine. "It is evident to every thinking person that it is impracticable to reduce the hours of labor in a manner whereby the world's product is curtailed or lessened in the least." Human advancement was based upon an increase in the production of material goods. Thus any reduction in hours had to be proportionate to an increase in production. Campbell pointed out that such had been the case in past reductions of hours.[24]

The Commissioner of Labor Statistics for New Jersey, James Bishop, reluctantly accepted the argument that "where machinery is very largely used" in the manufacturing process, shorter hours might mean a lessening of production. However, in other occupations, eight hours would produce as much as ten.[25] Other commissioners specifically denied that shorter hours would reduce production even in mechanized trades. Carroll Wright claimed that a reduction of hours for women in the textile mills of Massachusetts had not reduced production, but may even have increased it. Shorter hours led employers to use the existing machinery more efficiently or stimulated improved equipment. In addition, management was more careful, and "stricter discipline" was enforced to maintain production at maximum levels.[26]

Although most employers continued to deny that production could be maintained in a shorter day, there were some firms that had reduced hours and suffered no loss of production. William Gray, Treasurer of the Atlantic Cotton Mills of Lowell, Massachusetts, pointed out that his company had reduced its ten and three-quarter hour work day to ten in 1867. Initially, this had meant an increase of just under 3 percent in the cost of labor and a reduction in production of from 4 to 5 percent. However, the shorter day had been accompanied by the increased use of piecework and by an increase of 4 percent in the speed of the looms. Three and one-half years later, production was fully equal to the old ten and three-quarter hour day, and this was accomplished without any improvement in machinery. Gray believed the same had been true earlier in England when hours for women were reduced from twelve or more to ten. One English manufacturer had reported to him that within five years production in ten hours equalled that formerly done in twelve, as a result of a shift to piecework, improvements in machinery, and more efficient work from the employees.[27]

The United States Armory at Springfield, Massachusetts, had been placed on eight hours in 1868, under a law passed by Congress that established a shorter day for all mechanics and laborers employed by the federal government. By 1872, the workers were producing so much more per hour that the management of the armory cut piece rates 16 to 17 percent and workers still earned what had once been the customary wage for a ten-hour day.[28]

Throughout the late nineteenth century, labor leaders continued to insist that shorter hours, accompanied by increasing productivity, would do nothing to absorb the unemployed.[29] The same number of workers would produce in eight hours what had formerly been manufactured in ten. Yet the absorption of the unemployed was the very heart of the appeal of eight hours to the labor movement. Full employment made possible the favorable supply-and-demand relationship for labor that trade unions believed was the basis for higher wages. It eliminated the excess of labor that so weakened the efforts to organize workers, and that placed such a brake on the power of labor organizations.

In addition, the attempt to increase productivity per man-hour seemed to invite the speedup. To the union leader, an eight-hour day was of little value if workers were to be exposed to the tensions of a piecework system, which was manipulated so that workers had to increase the amount of labor and production necessary to earn a given wage. Since one objective of eight hours was to provide more leisure, it was self-defeating to accept shorter hours but harder labor, so that workers would be as exhausted in eight hours as they had been in ten.

In the effort to win a shorter day, labor leaders not only faced the hostility of most employers and the refusal of reformers to back a restriction of productivity, but considerable resistance from within their own ranks. It was widely argued, both inside and outside the labor movement, that the reduction from ten to eight hours would mean a corresponding loss of wages. The rejection of increased productivity per man fortified such a conclusion, since reformers who favored a shorter work day insisted that workingmen could not expect to produce less and receive the same wage.[30] Although labor leaders denied that wages would fall, they could not discount the strength of the general belief among workmen that such would be the result.[31] Workers already faced privation with the earnings of the longer day: "Men say, if they work less they get less, that they must work all they can to live."[32]

In 1871, the Cincinnati local of the coopers' union threatened to withdraw from the organization because the national convention insisted on a reduction in hours to ten per day. The local feared that the shorter day would further diminish their already inadequate wages.[33] Despite the long history of support for eight hours within the Machinists' and Blacksmiths' National Union, John Fehrenbatch, president of the union, found it necessary to remind workers in 1875 that they were responsible for the lack of progress in achieving eight hours. Why should employers concede eight hours "when the great body of workingmen prefer the ten hour rule."[34] Ira Steward also recognized the importance of the fear of a wage reduction in blocking shorter hours. He believed only practical experience with shorter hours would convince workers that their wages would not decline. Thus the importance of even limited experiments with the shorter day, for these would provide the necessary evidence that eight hours did not mean lower wages.[35]

Supporters of eight hours argued that the workers' fear of a wage loss was groundless. The reduction of hours during the nineteenth century, from approximately fourteen per day to ten, had been accompanied by wage increases, and the English experience with the ten hour law for women and children in 1847 indicated no linkage between shorter hours and lower wages.[36] The same was true where differences in hours existed within an industry. As the result of a state law, the textile mills of Massachusetts operated on a sixty-hour week while mills in New York and the other New England states worked sixty-five or sixty-six hours per week. Yet wages in Massachusetts were higher than in any of the other states except Rhode Island.[37]

In 1889, the Maine Bureau of Labor Statistics investigated the relationship between hours and wages among unionized granite cutters—a skilled, hand occupation—and found much the same situation as in the highly mechanized textile industry. Unionized granite cutters in the United States worked from forty-eight to sixty hours per week, yet average daily wages were not lower where hours were fewest. Fifty-eight union locals worked fifty-eight to sixty hours per week at an average daily wage of from $2.87 to $2.90 per day. Two locals had gained a forty-eight hour week at an average daily wage of $2.91 while thirty-three locals worked fifty-three to fifty-four hours per week at average daily wages of from $3.33 to $3.55 per day.[38]

Labor leaders insisted that such results would be typical, as hours were reduced, because wages reflected the supply-and-demand situation for labor, and shorter hours affected this crucial factor in favor of the worker. The merits of individuals could not long withstand the supply-and-demand pressure on wages, and even without the additional threat of machinery, a molder in 1863 asked whether there was "a single mechanic who has worked five years at his business who has not seen cases where first class workmen were discharged, and miserable botches hired in their stead, for daring to ask an advance of wages."[39]

Organized labor firmly believed that wages were ultimately set by what the idle man would accept.[40] Any factor, such as machinery, that created a permanent pool of idle workers exerted a downward pressure on wages.[41] Trade unions had to restrict the supply of men so that it was equal to, or less than, the demand. Apprenticeship regulations, the restriction of immigration, and shorter hours all worked to this end. Eight hours would absorb the idle and would contribute to the trade union's control of the labor market through a better balance between the supply of labor and the demand.[42]

Once such a balance was gained, wages would rise for two reasons. First, the supply of labor would be less than, or only equal to, the demand, and thus ordinary market conditions would be more favorable to the worker.[43] In addition, the absorption of idle workers would greatly strengthen the union's ability to take advantage of the more favorable market situation in its bargaining with employers. Why were so many strikes lost? The labor leader and anarchist A. R. Parsons believed the problem was basically a result of the adverse market for labor. Commenting on the printers' strike against a wage reduction at the *Chicago Inter-Ocean* in 1878, he pointed out that "Every man in the office, except one, left at the command of his union, but as *they* went out the front door the *hungry* 'scabs' entered at the back. . . . Labor will continue to suffer defeat until it learns how to take its surplus from off the market by reducing the hours of labor until there are no unemployed men."[44]

Despite the campaigns for the eight-hour day, the problem of convincing workers to back shorter hours continued. The labor editor and leader of the Knights of Labor, Joseph Buchanan, argued in 1891 that union members would oppose employers only on wage questions. In many trades, the failure to win eight hours had resulted from the refusal of large numbers of rank-and-file members to fight for it.[45] In the 1890's, supporters of eight hours still found it necessary to argue against the "generally" held belief among workers that shorter hours would mean lower wages.[46]

Although the Knights of Labor supported the eight-hour day, the organization took few concrete steps to secure it.[47] Terence V. Powderly, Grand Master Workman of the Knights, denied that an eight-hour day, without cooperative control of machinery, could be anything but a palliative for the worker. Powderly denied that shorter hours would permanently reduce the pool of unemployed. Employers would quickly erect more machines and employ more women and children to operate them in place of men. Within a short time, workers would find themselves faced with the same problems despite shorter hours. Restating one of his basic themes, Powderly argued that mechanization had to be met by the unity of the artisan and the laborer. Otherwise the displacement of skilled workers would lead the $3 per day man to compete with the $1 per day laborer for work. Unity among workers would lead eventually to cooperative control of machinery as part of the society that would supersede the existing capitalist system.[48]

The Committee on the Eight Hour Day reported to the convention of the American Federation of Labor in 1898 that workers continued to require education on the need to limit hours. "Selfishness and fear of recurring depressions, with their suffering and poverty, combine to make men eager to work to the limit of endurance when opportunity affords. . . ." The committee called on the A. F. L. to undertake a more active educational campaign for eight hours through literature, the labor press, and public speeches, while local unions were urged to set aside a portion of each meeting for discussion of the hours question.[49] That such a suggestion had

to be made, after more than three decades of unrelenting publicity in support of the eight-hour day, indicates the strength of the rank and file's
reluctance to embrace the shorter work day.

By the end of the nineteenth century, only a small portion of America's
unionized workers had gained the eight-hour day.[50] Workers observed that
employers resisted a shorter day, without a wage reduction, more strenuously
than they did demands for wage increases without any change in hours. Employers demanded wage reductions or the introduction of piecework—with
its possibilities for the speedup—in exchange for any reduction of hours.
Workers in weaker unions were thus more inclined to follow the line of
least resistance and press for wage gains rather than a reduction in hours.
Only unions that had strong control of their trades could win the shorter
day and make it stick. In fact, only the strongest locals, in the best-organized
building trades, had gained the eight-hour day by the 1890's.[51]

Employers opposed the shorter day more stoutly than wage demands
for much the same reason that they often refused to recognize unions while
making wage agreements with them. Union recognition and shorter hours
made a more permanent impact on the employer's freedom of operation
than agreements on wages. Wages constantly changed, and industrialization
created conditions which undermined older factors that had given skilled
workers considerable bargaining power. A higher wage in a prosperous year
did not prevent a wage reduction should business conditions or the labor
market change. However, a reduction in hours was more permanent. The
history of the nineteenth century revealed that once hours were reduced
they became a customary standard that was rarely increased. Thus a reduction in hours had to be regarded as a fundamental change that would affect
other aspects of the business.

Union recognition raised the problem even more starkly. It meant that
freedom of operation by management was now limited to a greater or
lesser degree, a condition that most employers resisted as strongly as possible. Accordingly, most strikes in the late nineteenth century were over
wages, where the issues of permanence and impingement upon control of
the business by management were least important. The most bitter strikes,
which were often lost by unions, concerned demands for more fundamental
changes, including reduction of hours.[52]

Employers also resisted eight hours because they were convinced that the
theory behind the movement for shorter hours might be correct. A reduction in hours, without a corresponding increase in productivity, might well
soak up a goodly portion of the unemployed, which would improve the
bargaining power of workers and their unions. Acceptance of the overproduction thesis, with its insistence that limits on productivity were socially
useful, promised to impede mechanization. Both these developments could

limit profits in a significant way. Eight hours thus meant more than a reduction in the work day. It carried implications that were potentially dangerous to the freely developing American industrial economy, and it further threatened to increase the power of the trade union, which was one of the few opponents of such unrestrained industrialization.

In practice, the opposition of employers was crucial, and many unions pulled back from the bitter struggle that would have been necessary to win the shorter day. Instead they concentrated on the more achievable gains in wages. Despite the great appeal of shorter hours, only small groups of workers won the eight-hour day on the basis of the labor movement's position on productivity. Most American workers gained the shorter work day in the twentieth century by accepting increased productivity. Thus trade unions generally failed to gain shorter hours on their terms, significantly weakening their efforts to meet the effects of industrial change.

## CHAPTER TEN

# IMMIGRATION AND INDUSTRIAL CHANGE

Organized labor's insistence on apprenticeship rules, its strong support of laws to control the hours of women and children, and the emphasis on the shorter work day were designed to contain the impact of new methods of production upon the existing work force. However, these measures affected only workers already in the United States. Another avenue of response was to limit newcomers to the country, since they might also serve as agents of industrial change.

American workers had always been hostile to competition from immigrants, but industrialization heightened the fear of the newcomer by intensifying that competition. As changes in the methods of production reduced the protection of skill, immigrants could more easily challenge those already at work. It was not the traditional competition of a newly arrived artisan and an American skilled worker. Instead the immigrant often became the machine operator, or the "botch" workman, who quickly learned the simpler tasks as production was divided. The availability of female, child, and immigrant labor for the low-wage positions as machine tenders encouraged employers to abandon older hand methods of production. The immigrant thus became a human factor in the process that displaced skill itself.

Organized labor consistently opposed immigration in one form or another. The labor movement did not move from support of immigration, with exceptions, to opposition to immigration, with exceptions. Rather it consistently opposed those forms of immigration that seemed a direct threat to the skilled labor force at any particular time. Initially, organized labor opposed the "importation" of new workers, which meant all direct efforts by private or governmental means to stimulate foreign workers to emigrate to the United States. Contract labor was one specific form of such importation. When these efforts failed to solve the problems associated with immigration, the labor movement shifted toward a more complete restriction in the late 1880's and 1890's. The magnitude of immigration, the effect of industrialization in bringing such newcomers into competition with the existing labor force, and the growth of a sentiment for restriction from outside labor, which made restriction on a wide scale politically viable, ultimately led organized labor to adopt a broad exclusionary policy.

The labor movement of the 1860's sought to stop the importation of foreign workers. One major step to this end was the end of contract labor; another the cessation of Chinese immigration, which labor leaders regarded as a clear example of the importation of an entire national group. Labor leaders believed that immigrants under contract were recruited specifically to break strikes and undercut wages. In 1864, contract labor had been the means by which Giles F. Filley, of St. Louis, hoped to avoid the use of union men in his foundry. Forty-two molders were brought from Prussia, on contract, even though union men were freely available. The Iron Molders' International Union met this challenge by contacting the men and providing for them "until they could secure work upon honorable terms." This was not an isolated case, for over two hundred molders had been induced to come to the United States from England, Scotland, and Germany by promises of steady employment at good wages. At the same time, many American molders were out of work. The Molders' Union recognized that the newcomers were poor persons in a strange land, and though they were honorable men who would not work once they learned the true situation, the Union nonetheless felt "obliged to keep them until they could procure employment."[1]

In 1867, workers in Allegheny County, Pennsylvania, issued an address which stated "That the purchased importations of foreign labor by employers for the purpose of reducing the price of labor shall, as far as practicabie, be prevented, but, all *voluntary* immigrants from every country of the globe shall be heartily welcomed to our shores."[2] The National Forge of the United Sons of Vulcan also attacked contract labor, but proclaimed "this land the asylum of the oppressed of all nations," who were invited as voluntary immigrants. The invitation, however, did not include the Chinese, who were specifically excluded.[3] Other unions also endorsed voluntary immigration while attacking contract labor in general or Chinese immigration in particular.[4]

By snapping the direct link between the immigrant and the employer, through the end of contract labor, organized workers believed that the threat from the newcomer would be greatly reduced. If skilled immigrants came freely, they were less likely to be available to an employer intent on avoiding the use of union men, they would be more open to contacts with American unions, and they would be distributed over a wider area, so that their impact upon any one place would be reduced. Trade union leaders believed that many of the skilled immigrants could be organized. S. R. Gaul, President of the Bricklayers' National Union, recommended in 1869 that immigrant bricklayers with cards from European unions be admitted to American locals without initiation fees. It was essential that they be organized immediately; otherwise they accepted work for less than the union scale, which injured the interests of all workers in the trade.[5]

A limited number of immigrants could be absorbed into the labor market without an adverse effect on American workers. However, trade union leaders believed that employers planned to use labor agencies to flood the American market with great numbers of European workers, over and beyond those who would come freely. Jonathan Fincher pointed out that this vast increase would produce disastrous results. American workers had escaped the poorest conditions found in the Old World because they lived in a young nation with a limited population. The result of the wholesale importation of labor would be twofold: the wages paid workers would drop, and the increased production would lead to glut and unemployment. Fincher thus concluded that labor's self-interest required determined opposition to such schemes.[6]

The labor movement opposed not only workers who came under specific contracts, but the much larger group who were supposedly stimulated to emigrate by misinformation concerning labor conditions in America. For labor leaders, this propaganda for emigration was a more potent encouragement to Europeans than the traditional lure of the New World or the advertising of shipping companies. They opposed the activities of private organizations, such as the American Emigrant Aid Society, which clearly aimed to serve the interests of employers. However, organized labor also opposed the use of American consulates to encourage immigration. Advertising for newcomers by the state or the federal governments also was regarded as an artificial inducement. All these efforts were considered steps toward a massive importation of new workers.

For the American labor movement, the crucial distinction was between free, voluntary immigration and importation—even without specific contracts. Imported workers usually lacked the normal network of contacts that encouraged emigration. By the 1860's newly arrived skilled workers from Europe often had friends or relatives in the United States, including members of labor unions. Some of these immigrants had been union members in their home countries. In contrast, imported labor was regarded as artificially stimulated immigration, and the workers who arrived were more isolated, which made it less likely that they would demand American standards of wages or join trade unions.

In an effort to offset the activities of the agencies that stimulated a desire among European workers to emigrate to the United States, American labor leaders sought the help of European trade unions, particularly those in Great Britain. The European trade unions could counteract propaganda about rosy conditions in America, point out that the aim of the American Emigration Aid Society was to overstock the American labor market to the detriment of the newcomer as well as the established worker, and could caution against the use of contract labor as a source of strikebreakers or as an alternative to union men. To the chagrin of American labor leaders, little cooperation was received. William Sylvis, President of the Iron Molders' International Union, contacted leaders of the iron molders in Scotland and England concerning immigration; but he concluded sadly that "Up to the present time, their action, or rather their want of action, would seem to indicate an entire indifference on their part and a greater confidence in the agents of the American Emigrant Society than in the officers of the Iron Molders' International Union."[7]

Jonathan Fincher believed that this inaction was motivated by the desire of Scotch and English trade unions to arrange for the emigration to America "of vast numbers of their members as a ready means of obtaining control of their trades. . . ."[8] In 1867, the National Labor Union continued the attempts to gain the cooperation of European trade unions.[9] However, the emphasis soon shifted to curbing the efforts of the state and federal governments in support of increased immigration. In 1868, the National Labor Union contended that the objective of these activities was "to control the price of human labor" through bringing foreign workers into "direct competition with American labor."[10]

Chinese immigrants were regarded as the clearest example of imported labor, and the threat that such immigration posed became personified in the Chinese worker. American labor leaders charged that Chinese immigrants were "imported by capitalists in large numbers, for servile use. . . ."[11] Accordingly, their wages were set far below existing levels. Labor leaders also stressed that the Chinese were isolated from other workers: they did not generally become citizens, lived within their own communities, and sent much of the proceeds of their labor to China. As a result, the labor movement concluded that the Chinese were beyond the influence of organized workers and available to serve the employers' efforts to depress wages in favor of increased profits.[12]

The few efforts to use Chinese workers in the Eastern United States excited intense fear among labor leaders, even though the number of Chinese workers was small. The issue was not numbers, but the possibility that the success of these limited efforts could lead employers to import larger numbers of Chinese into the Eastern United States.[13] Outside the West Coast, the threat was thus more potential than actual, but quite real for labor leaders nonetheless.

In response to organized labor's attacks on Chinese immigration, the Labor Committee of the New England Shoe and Leather Dealers' Association argued that immigration and machinery had the same purpose: a diminution of the cost of production. The "cheap labor furnished by the immigrant" was beneficial to the entire community. Progress had to be measured in what a day's work would buy, and lower costs of production were the key to such an increase in real income. As one employer put it: "If it [Chinese labor] is objected to because they work at less price than our people, the objection exists with ten-fold more force against the sewing and pegging machines . . . and before you stop Chinese immigration, you should abolish the sewing and pegging machines."[14]

Racial slurs were a prominent part of the campaign against the Chinese. One could easily cite the attacks in labor sources on the Chinese as pagans, as immoral, bestial, and filthy in their habits, and as "rat eaters," debased by centuries of the most abject misery and serfdom.[15] However, racial hostility toward the Chinese was not the basic reason for the campaigns to bar their further entrance into the country. Instead, racism fortified the fear of economic competition and allowed for its most emotional expression.

There would have been little love for the Chinese had they come as free immigrants, for ethnic and racial hostility, based upon the presumed superiority of one's own group, was a basic feature of American society in the nineteenth century. It fueled the hatred of the Irish before the Civil War as fully as the attacks on the southern and eastern Europeans in the 1880's and 1890's. The Chinese felt the sting of this hatred and animosity as well, but this they shared with many other immigrant groups.[16] What led American labor to convert its private dislike of the Chinese, its private racism toward the Chinese, into an organized effort to exclude them from the nation was concisely spelled out by the Protestant minister Joseph Thompson in 1879. He wrote that the opposition of American workers to the Chinese was not because of "what they eat, what they wear, what they believe, but that they compete with other workmen by cheap labor." Thompson added that if the Welsh, Poles, or Icelanders offered to work cheaply, there would be the same opposition to them.[17]

American labor leaders demanded the end of all Chinese immigration because they saw little possibility that Chinese immigrants would be anything but cheap labor. The representative of the Central Labor Union of New York City, Louis Post, appeared before the Senate Education and Labor Committee in 1883. He testified that the Central Labor Union had opposed

Chinese immigration because it was an example of imported cheap labor, not because the workers were Chinese. Free immigrants were still welcome, and had the Chinese been of this type, they would have been accepted. When asked whether any part of the opposition to the Chinese was "based on the supposition that the race had been inured to a hard, low manner of living and to low wages so that there is an indisposition on the part of American workingmen to come in competition with labor of that class," Post replied that a feeling of that kind was quite general among American workers.[18]

In 1894, Samuel Gompers condemned the "servile of all nations, our own included," who undercut American standards by acting as cheap, docile labor. He believed that the Chinese were clearly servile labor, since their experience in China had been one of submission. Gompers placed the blame clearly upon the oppressed as he chastised the Chinese for allowing themselves to be so "barbarously tyrannized over in their own country."[19] Thus the Chinese were likely to be cheap labor in America because they had been cheap labor in China.

The belief that Chinese immigrants would not overcome their initial acceptance of low wages was basic to the argument for exclusion. The editor of the *Cigar Makers' Journal* insisted in 1894 that the hostility to the Chinese was not a question of sentiment or racial prejudice. Instead, it was a matter of economic competition. "We simply cannot compete with a Chinaman, and we have no use for that line of statesmanship that will bring the standard of living of the American working men down to a Chinese level. . . . If, however, the Chinamen would adopt our ways and customs, there would be less logic in our position, but as a matter of fact, he will not."[20]

Henry George agreed that the Chinese would not adopt American ways. But he viewed assimilation as more a matter of culture than will. The Chinese were not an inferior race, but simply a vastly different one whose customs made assimilation unlikely. European immigrants were also cheap labor upon arrival, but they were close enough to the American culture to be assimilated in time. Once incorporated into the American nation, they would demand the existing wage scale; but so long as an immigrant did not assimilate, he would continue to act as cheap labor.[21]

In fact, there was little opportunity to test the adaptability of the Chinese immigrant, since the conditions of his entry, along with the hostility from American workers, combined to isolate him and make any real contact with American labor unlikely. However, the American labor movement assumed that such assimilation was impossible, and thus labor leaders believed there was no way to combat the economic competition from the Chinese immigrant except by preventing his entry.

The imported labor theme never was restricted only to those immigrants who could be truly classified as having entered under some form of contract. It always rested upon the broader concept that certain immigrants were more dangerous to the interests of American workers because of the specific conditions of their entry and their subsequent behavior in the United States. Thus French-Canadian immigrants found themselves under attack in a manner quite similar to the Chinese.

French Canadians had emigrated to the United States since the 1840's. Many worked in the industrial towns of New England, and, by the 1870's, French-Canadian women and children were a major element in the labor force of the mills. The men were employed in a wider range of occupations but often had lower wages than the existing labor force. Unlike the Irish, French Canadians often worked in the United States only during the busy season in the mills. To American labor leaders, this meant that the French Canadians would have little interest in the uplift of conditions within the United States. Instead, labor leaders believed the objective of the French Canadians was the short-run accumulation of income, with the intention of returning temporarily or permanently to their original homes north of the border.

In 1869, this situation led E. L. Rosemond, Secretary of the Carpenters' and Joiners' National Union, to castigate the French Canadians in terms similar to the attacks upon Chinese immigrants.

They have no interest in our cause or country. They are not citizens, nor do they intend to be. Still these men are often employed in preference to actual residents. While our sympathies should extend to the oppressed of all nations, and the right hand of fellowship be given to those who come among us with the intention of adopting our country as their home, on the other hand he who comes among us for the purpose of taking his earnings to a foreign clime is a moth to society and should be scouted from our midst.[22]

For American workers, the Chinese were cheap labor and likely to remain so, because they were imported workers whose habits and customs isolated them from the existing work force. The French Canadians were also considered cheap labor and likely to remain so, because they had no permanent interest in settling in the United States and thus were interested only in the income that could be earned in the short term.

The heightened campaign against the Chinese that led to the suspension of immigration in 1882 also provoked a new series of attacks on the French Canadians as well. Commissioner of Labor Statistics for Massachusetts, Carroll Wright, used the pages of the annual report to characterize the French Canadian as "the Chinese of the Eastern States." Wright noted "some exceptions," but generally the French Canadians cared nothing for American institutions because they were not permanent residents. Instead they worked for a few years in the American mills, at whatever wages they could get, saved as much of their earnings as possible, and ultimately returned to Canada. They were utterly indifferent to schools for their children, preferring to have them work in the mills. Their personal interests seemed to be "drinking and smoking and lounging" and little else. In the manner of the Chinese, "they are a horde of industrial invaders, not a stream of stable settlers."[23]

These comments by Wright provoked a strong defense from within the French-Canadian community, and Wright ultimately admitted that his statements, based upon the material provided by "informants of the Bureau," were too sweeping.[24] However, the comparison of the French Canadians and the Chinese touched a sympathetic chord among labor leaders who faced competition from French-Canadian workers. Thus the labor editor George Gunton defended Wright's characterization of the French Canadians as a simple statement of "what all who work or live with them see and believe every day. The people have not endorsed Mr. Wright's sentiments, but Mr. Wright has endorsed the sentiments of the people. The same features that make the Chinaman objectionable apply to the French Canadian also but with less force." Gunton specified that the Chinese and French Canadians were similar in maintaining separate languages and religions; in remaining aloof from American society, huddled together "in unwholesome numbers in the very poorest tenements"; in accepting the lowest wages for any numbers of hours; and in their docility to the wishes of the employer.

However, Gunton refused to follow the path of many labor leaders and assign these negative characteristics to ethnic or racial qualities. Instead, he argued that this behavior was not peculiar to any ethnic group. "People don't live on rice and rats, work long hours for low wages, huddle in filthy tenements in the back slums, and ransack ash barrels for a living because they are Chinese or French or English or Irish or Dutch or Italian or Turk, but because they are poor." The Chinese accepted the status of imported labor because they were poor; the French Canadians acted as strikebreakers and accepted intolerable wages and tenement conditions because they were poor.

The rich Chinese don't live on rice and rats and bunk on a board. The rich Frenchmen don't live on six-and-a-half-street. The rich Irish don't crowd the steerage quarters of Atlantic steamers and fill the tenement houses of the large cities. The rich Italians don't forage ash barrels for a living. No, these people do these things not because of their nationality or religion but because of their poverty. This is not a race but a poverty question.

Gunton then called on the French Canadians to stop denying these conditions and join labor and reformers in alleviating the situation through compulsory education for children, and ten hour laws for women employed in factories.[25] Many labor leaders believed that the more effective course was to restrict the entry of the poor rather than correct the conditions produced by their arrival.

Chinese immigrants were clearly not a unique threat to American workers, and the assault upon them was not marked by racist arguments that were omitted in the case of Europeans. Whenever an immigrant group offered severe economic competition to American workers, the hostility engendered eventually was expressed in terms of racial and ethnic inferiority. Thus the journal of the skilled shoe lasters strongly supported the exclusion of "degraded" European immigrants. Among those to be barred were Hungarians, Armenians, and Russian Jews. These immigrants were in the same category as the Chinese since they were lawless (and slavish as well in the case of the Armenians) and formed clannish ghettos. Yet the president of the Lasters' Protective Union made it clear that these undesirable personal and ethnic characteristics were not the basic issue. Instead the threat to American workers came from the willingness of certain European and Asian immigrants to retain the standards of their homelands and to work at any price. He believed that all such immigrants should be barred from the United States.[26]

The Chinese were singled out for exclusion first because they seemed to constitute the clearest threat, while their small numbers and geographic concentration meant that they were powerless to defend themselves as a group. In addition, they were of little real importance to most of American industry. Thus employers could accept the closing of Chinese immigration in 1882 since the flow of labor from southern and eastern Europe was increasing. Organized labor could be given the illusion of legislative victories in the suspension of Chinese immigration in 1882 and the abolition of contract labor in 1884, since neither of these sources of immigrants was of real importance. Most newcomers that entered the United States were free immigrants from Europe, and neither law would have impact on this basic source of labor.

As industrialization made the threat of the European immigrant all the more dangerous to American labor by destroying the protection once offered through skill, and as the number of immigrants increased rapidly in the 1880's, organized labor moved beyond the laws of 1882 and 1884 toward a policy of broader restriction. In the campaign for the end of contract labor that culminated in the Foran Law of 1884, labor spokesmen still stressed that many workers would never have come to America but for the inducements of employers.[27] However, Peter J. McGuire, founder of the

United Brotherhood of Carpenters and Joiners, recognized that free immigration would continue to offer a competitive threat to American workers, even were contract labor ended. If this were true, "why should you not prevent the abnormal increase of that competition."[28]

As the Legislative Committee of the Federation of Organized Trades and Labor Unions reported to the convention in 1885, the contract labor law of the previous year might be evaded, and even if enforced, might bar only a small number of immigrants. Yet it was still a step in the right direction.[29] Thus even as contract labor was barred by law, labor leaders recognized that it was no longer the major threat. The end of contract labor did not slow the vast increase in immigration, for the increased number of newcomers was basically a response to the growth of an American industrial economy, not the efforts of labor agents.

Not only did the end of contract labor fail to reduce the number of immigrants, but even more important, it had no effect in lessening the impact of the newcomers upon the standards of American workers. It became increasingly apparent to labor leaders that educability was not a property of free or contract labor as such. Free immigrants also rejected the existing standards of American workers and thus provided much of the same type of competition previously ascribed solely to imported labor. The situation was made all the worse by the vast increase in the number of immigrants despite the Foran Law. Thus trade unions moved from the earlier emphasis on the end of imported labor to a broader policy of restriction, which promised much more substantial limits on the total number of immigrants as well as a heavier impact upon the Southern and Eastern Europeans, who were regarded as less assimilable and less open to unionization. Accordingly, a policy of restriction, based on literacy and means, replaced the older criteria of importation as against free entry.

Although the American Federation of Labor did not endorse the literacy test as the basis for admission into the United States until 1897, the acceptance of a broadly based policy of restriction had wide support among American workers throughout the 1880's and early 1890's. The state bureaus of labor statistics reflected the concern of workers about immigration, and discussion of the issue in their reports increased dramatically. The sentiments of workers and commissioners were strongly anti-immigration, with little distinction between imported labor and other types of entry.[30] Although there were some attacks on the newcomer because of his personal habits or beliefs, the basic thrust of the demands for restriction was the economic threat offered by the immigrant. As one teamster put it in 1885: "I am not opposed to a man because he is a foreigner, but I think there are too many here for the work that [there] is to do."[31]

It was also pointed out that much of the pressure for immigration came not from good conditions in the United States as much as from poor conditions in Europe. Thus unlimited immigration meant that the United States was attempting to solve problems of unemployment and poverty for much of Europe as well as for itself.[32] John Coffey, President of District Assembly 149 of the Knights of Labor, acknowledged that conditions for glass blowers were much poorer in England than in the United States. He attributed this in part to the emigration of Swedish glass workers to Great Britain, and thus he was not prepared to suffer the same situation in the United States as the result of the emigration of British glass workers to this country. Although sympathetic to the European mechanic, Coffey urged them to organize and fight at home to win better conditions.[33]

Continued immigration could also offset the positive effects that were anticipated from the eight-hour day. The focus of the movement for shorter hours was to reduce the oversupply of labor, to the benefit of American workers. However, immigration added vast numbers of new workers, which offset the positive results of the shorter work day.[34] The early critic of unrestrained industrialization W. Godwin Moody argued that in a time of incomplete employment, "America might very properly be preserved for the Americans"—a sentiment that won increased support from organized labor through the 1880's and 1890's.[35]

In 1886, a popular treatment of labor made the connection between industrial change and immigration crystal clear.

It is the development of machinery, of labor-saving devices, which has interfered with the benefits which immigration, up to a certain point, undoubtedly has produced. Our home-made surplus population is increasing annually in numbers, as it must under our industrial conditions. Labor-saving machinery, it is estimated, annually displaces 70,000 workers, who are thrown on the street to find some avenue of work. To these must be added the immigrants who also annually come to our shores. There are honest, industrious, intelligent and effective men who cannot find work in the United States. The supply of labor is already greater than the demand, and yet we go on adding to the supply by every steamer which crosses the Atlantic, and subtracting from the demand by every patent filed in the patent office.[36]

Trade unions increasingly accepted this type of analysis. They viewed immigration, in all its forms, as one more element in the negative effects of unrestrained industrialization.

The immigrant had always been a competitor, but in the late nineteenth century that competition became more ominous as the lessened importance of skill and the presence of the unemployed breached labor's defenses. Hours had to be reduced, the labor of women and children had to be curbed, the number of apprentices had to be limited, and immigration had to be restricted. It was a total response, a whole cloth, in which the attitude towards immigrants was intertwined with the other strands of the fabric. No one step was a panacea. Even were all the desired restrictions on cheap labor won, there would still remain the problem of migration from American rural areas, as well as the willingness of those already in the labor force to undercut each other in the competitive labor market. However, to allow massive additions of immigrants to the American work force seemed folly to the labor movement. Seemingly, immigration and industrialization became antithetical for the American labor movement.

The American Federation of Labor considered an educational test for immigrants at its convention in 1896. It was designed to reduce the total number of immigrants as well as eliminate many of the poorer newcomers, who were considered the greatest threat to American workers. The Special Committee on Immigration made it clear that they were not anti-immigrant in any spirit of know-nothingism, which would use the question of immigration "as a pretext to gloss over social wrongs." Instead it was a question of the existing economic conditions that consigned "so many willing workers to idleness. . . ."[37]

In 1897, the A. F. L. endorsed immigration restriction based upon an educational requirement and a means test by a vote of 47 to 18. The proposal envisaged a literacy test for all immigrants over fifteen years of age (except for wives or aged parents of persons already residing in the United States) who were without sufficient means for immediate self-support. The delegates who offered the resolution made it clear that immigrants who were illiterate and poor were more likely to gravitate to urban industrial areas, where they became victims of unfair employers.[38] Clearly the purpose was to shut out the poorest immigrants in an effort to reduce the likelihood that immigrants would accept the lowest wage offered in order to maintain themselves.[39] The old emphasis on eliminating a specific type of immigration, such as contract labor, was no longer a viable protection for American labor, "there being little occasion for employers to contract for labor abroad when workmen can be so easily induced to come. . . ."[40]

By the time the American Federation of Labor adopted the combined literacy and means test for immigrants, important elements within organized labor were already pressing for even more restrictive action. As early as 1889, a resolution was presented to the convention of the A. F. L. calling for a complete suspension of immigration for fifteen years. Although this proposal was not endorsed by the convention, it received thirteen favorable votes as against the twenty-eight for referral to the Executive Council.[41]

In 1896, the convention of the United Brotherhood of Carpenters and Joiners demanded not only the literacy test but additional legislation limiting the number of newcomers, regardless of literacy, to 50,000 per year, and establishing a board drawn from various labor organizations which "shall have the power to distribute immigrants to localities that are not already overcrowded with laborers. . . ." Those immigrants who could not find employment in the selected localities during a sixty-day period would be returned to their country of origin.[42]

Other labor organizations also thought the literacy test too mild a restriction. The New York state branch of the A. F. L. conducted a referendum among its members on the restriction of immigration. Acting upon the results, the state organization called for the suspension of all immigration for five years.[43]

Despite the opposition to restriction from newly arrived immigrants within the labor movement, who feared restriction would exclude their fellow countrymen; despite the opposition of many socialists, who identified the enemy as the capitalist system, not the immigrant, and who felt that restriction only intensified the hostilities among different ethnic groups within the labor force, to the advantage of employers; and despite labor leaders, who continued to argue that "What was needed in this country was the restriction of the machine and not of immigration,"[44] American organized labor turned to a broad restriction as one more means to combat the impact of industrialization. The proponents of a broad restriction were able to carry the day because they stressed the economic threat that workers believed was implicit in large-scale immigration.[45]

Ultimately American labor identified the immigrant with industrialization, and the newcomer came to personify the basic problem of competition among workers that seemed to threaten American labor. As the number of immigrants increased, industrialization reduced the traditional defenses and means of assimilation formerly relied upon by organized labor. Faced with rapid and extensive industrial change, the labor movement sought to cushion the shock of new conditions upon the established labor force. In the process, the aspirations of those from abroad could hold little weight in the face of the necessity to defend against unrestrained industrialization.

In an economic order that stressed the victory of competitive self-interest, the trade unions of the nineteenth century acted as they had to if they were to protect their members. This sharp clash of self-interest, this utter lack of a general good, or even a moderating agency, led the progressive reformers of the late nineteenth and early twentieth centuries to seek some means for rationalizing the economy so that the economic needs of society were not served at the expense of its members. Not until the 1930's, however, did the United States accept meaningful control of the economy in pursuit of some rational goals, and significantly curb the effects of unrestrained competitive capitalism.

In the last half of the nineteenth century, labor leaders could not await such developments, yet their response to the problems raised by industrialization proceeded toward the same end. Capitalism should remain, but it had to be decisively modified. Trade unions attempted to use work rules and agreements with employers, apprenticeship regulations, shorter hours, immigration restriction, and the control of female and child labor to interfere with the free operation of the new industrial economy. Such restraints stimulated the development of a system of state capitalism, in which the government actively regulated the economy. One major purpose of government action was to reduce the negative effects of uncontrolled struggle among organized economic interests. Economic conditions were no longer to be the result of market factors alone, but of political decisions that modified the market.

Stronger trade unions resisted government interference in their affairs, but this was only because they had already gained a measure of control over their industry sufficient to modify unrestrained industrial change. Weaker unions had no objection to regulation of the economy by the state. However, the basic aim of all trade unions was compatible with government regulation of the economy, since their objective was to eliminate labor costs from the uncontrolled operation of market forces. Although progressive reformers of the early twentieth century often attacked trade unions for concentrating on their own interests, the concept of the modern American system of state capitalism was dependent upon just the type of attack upon a free market that trade unions had mounted. Social Progressives would call for a new capitalism based upon social justice rather than craft interests and regulated by the government commission rather than the trade unions, but there was mutual agreement that industrialization had made it impossible for a modern economy to operate on an unrestrained competitive basis without destroying the interests of the community at large through the privation and poverty visited upon so many of its citizens.

# NOTES

## CHAPTER I

1. "Destiny of the Mechanic Arts," *Harper's New Monthly Magazine*, XIV (April 1857), 696–698.
2. National Typographical Union, *Proceedings*, 1858, p. 31.
3. Columns on scientific and technical advances were also found in other labor publications, including *Fincher's Trades' Review* and *The International Journal* of the Iron Molders' International Union.
4. *Workingman's Advocate*, July 11, 1868, p. 2; June 16, 1866, p. 2.
5. *Fincher's Trades' Review*, January 14, 1865, p. 28; Jonathan Grossman, *William Sylvis, Pioneer of American Labor: A Study of the Labor Movement During the Era of the Civil War* (New York, 1945), p. 143.
6. *Fincher's Trades' Review*, August 27, 1864, p. 50.
7. Ibid., July 29, 1865, p. 67. On this point, also see Richard Trevellick, ibid., May 27, 1865, p. 103.
8. Ibid., April 8, 1865, p. 74.
9. *Iron Molders' Journal*, XI (July 10, 1875), 355; (August 10, 1875), 392.
10. Emerson D. Fite, *Social and Industrial Conditions in the North During the Civil War* (New York, 1910), p. 197. Also see David Montgomery, *Beyond Equality: Labor and the Radical Republicans, 1862–1872* (New York, 1967), p. 105.
11. *Machinists' and Blacksmiths' International Journal*, IX (February 1872), 521.
12. For a discussion of the responses by the cigar makers and coopers to technological changes that threatened their skills, see Chapter V.
13. See Chapter VII.
14. *Fincher's Trades' Review*, February 20, 1864, p. 46; May 7, 1864, p. 90; May 13, 1865, p. 95.
15. James C. Sylvis (ed.), *The Life, Speeches, Labors and Essays of William H. Sylvis* (Philadelphia, 1872), p. 100.
16. *Fincher's Trades' Review*, June 27, 1863, p. 14.
17. See Sylvis's statement in *Workingman's Advocate*, November 2, 1867, p. 2, and the article by Fincher in *Welcome Workman*, February 22, 1868, p. 4. On immigration, see Chapter X.
18. *Fincher's Trades' Review*, July 25, 1863, p. 31.

19. Ibid., August 5, 1865, p. 76.
20. Ibid.
21. On Steward, see the cogent discussion in Montgomery, *Beyond Equality*, pp. 249-260. Also valuable is Dorothy Douglas, "Ira Steward on Consumption and Unemployment," *Journal of Political Economy*, XL (1932), 532-543. Born in 1831 and apparently self-taught, Steward was a machinist by trade. He was not only a theorist but aspired to be the leader of a national movement committed to the eight-hour day. During the 1860's and 1870's, Steward wrote and lectured in favor of eight hours, and he organized a succession of eight hour leagues. He died in 1883.
22. "Meaning of the Eight Hour Movement," Ira Steward Papers, Wisconsin State Historical Society, Box 3, p. 5.
23. See ibid. and "Less Hours" and "Unemployment" in Steward Papers, Box 3.
24. Massachusetts Bureau of Labor Statistics, *Fourth Annual Report*, 1873, p. 449.
25. *A Reduction of Hours an Increase of Wages*, 1865, as reprinted in John Commons et al., *A Documentary History of American Society* (Cleveland, 1910), IX, pp. 284-301. Also see Montgomery, *Beyond Equality* p. 254; Douglas, "Steward on Consumption and Unemployment," p. 535.
26. Untitled essay, Steward Papers, Box 3, p. 15.
27. "The Power of the Cheaper over the Dearer" as found in Commons, *Documentary History*, IX, pp. 309-310.
28. Thomas Phillips to Steward, Steward Papers, Box 3.
29. Untitled essay, Steward Papers, Box 3, pp. 2-3. Also see "Economy and Extravagance," Steward Papers, Box 3, p. 12.
30. "Wages and Wealth," p. 3; "Machinery and Wages," p. 2, Steward Papers, Box 3.
31. "Extravagance," Steward Papers, Box 3, p. 4.
32. "The Power of the Cheaper over the Dearer," Commons, *Documentary History*, IX, p. 313.
33. Ibid., pp. 314-325.
34. Ibid., p. 329; "Theory of Wages," Steward Papers, Box 3, p. 3.
35. "Theory of Wages," pp. 3-4; untitled essay, pp. 18-19, Steward Papers, Box 3.

CHAPTER II

1. On mechanization in the early 1870's, see Herbert Gutman, "Social and Economic Structure and Depression: American Labor in 1873 and 1874," Ph.D. dissertation, University of Wisconsin, 1959, pp. 203-240.
2. *New York Tribune*, May 28, 1872, p. 8.
3. Ohio Bureau of Labor Statistics, *Second Annual Report*, 1878, p. 275.
4. Open letter from Edward Rogers to Carroll Wright, *Labor Standard*, October 5, 1878, p. 5.
5. For examples, see *National Labor Tribune*, May 13, 1876, p. 2; *Labor Standard*, June 23, 1878, p. 1.
6. Connecticut Bureau of Labor Statistics, *First Annual Report*, 1874, p. 139.
7. *Labor Standard*, July 21, 1878, p. 1; July 28, 1878, p. 1.
8. *National Labor Tribune*, July 22, 1876, p. 2.
9. *Labor Standard*, November 2, 1878, p. 1. Also see October 21, 1877, p. 2.
10. United States Senate, Committee on Education and Labor, *Report upon the Relations Between Labor and Capital*, 4 vols. (Washington, D. C., 1885).
11. Executive Council, American Federation of Labor to the International Trade Union Congress, October 27, 1888, American Federation of Labor Papers, Wisconsin State Historical Society. Also see American Federation of Labor, *Proceedings*, 1897, pp. 18-19.
12. New York Bureau of Labor Statistics, *Eighth Annual Report*, 1890, pp. 693-702.

13.   Massachusetts, *Census of 1885*, 4 vols. in 3 (Boston, 1887–1888), p. cx;
Massachusetts Bureau of Labor Statistics, *Twenty-Fourth Annual Report*, 1893, p.
114.
14.   Massachusetts Bureau of Labor Statistics, *Eighteenth Annual Report*, 1887,
p. 294.
15.   Massachusetts Bureau of Labor Statistics, *Twenty-Fourth Annual Report*, 1893,
pp. 116–117.
16.   These examples are drawn from the tables on unemployment by occupation
found in the Massachusetts Bureau of Labor Statistics, *Eighteenth Annual Report*,
1887, pp. 280–283. A similar range appears in the figures for unemployment by
occupation among women. See pp. 283–284.
17.   Ibid., p. 3. Carroll Wright was the best known and most influential of the
commissioners of labor statistics. On his career and the activities and methods of the
various bureaus of labor statistics, see James Leiby, *Carroll Wright and Labor Reform*
(Cambridge, 1960).
18.   Carroll Wright, "The Relation of Production to Productive Capacity," *Forum*,
XXIV (February 1898), 665–666.
19.   Massachusetts Bureau of Labor Statistics, *Twenty-Fourth Annual Report*,
1893, p. 116.
20.   United States, Labor Bureau, *First Annual Report of the Commissioner of
Labor* (Washington, D. C., March 1886), pp. 80–90.
21.   Maryland Bureau of Industrial Statistics, *First Biennial Report*, 1884–1885,
pp. 143–144.
22.   Missouri Bureau of Labor Statistics, *Seventeenth Annual Report*, 1895, p. 95.
23.   Massachusetts Board to Investigate the Subject of the Unemployed, *Report*,
Part IV (Boston, 1895), pp. xlvii–xlviii.
24.   Ibid., p. xliv.
25.   Ibid., pp. xxxix–xl.
26.   Maryland Bureau of Industrial Statistics, *Third Annual Report*, 1894, p. 120.
27.   Maryland Bureau of Industrial Statistics, *Fifth Annual Report*, 1896, p. 43.
28.   Massachusetts Unemployed Board, *Report*, Part IV, pp. xxii–xxiv.
29.   Missouri Bureau of Labor Statistics, *Seventeenth Annual Report*, 1895, pp.
96–97.
30.   New York Bureau of Labor Statistics, *Thirteenth Annual Report*, 1895, pp.
370–372.
31.   Maryland Bureau of Industrial Statistics, *Fifth Annual Report*, 1896, p. 43;
Missouri Bureau of Labor Statistics, *Seventeenth Annual Report*, 1895, p. 96;
George Barnett, *Chapters on Machinery and Labor* (Cambridge, 1926), pp. 21–22.
32.   Barnett, *Machinery and Labor*, pp. 6–8.
33.   See Robert Ozanne, *A Century of Labor-Management Relations at McCormick
and International Harvester* (Madison, Wis., 1967).
34.   *The Carpenter* (November 1894), 4; (February 1899), 13.
35.   Ohio Bureau of Labor Statistics, *Thirteenth Annual Report*, 1889, p. 54.
36.   *Pattern Makers' Monthly Journal*, II (March 1893), 2.
37.   Massachusetts Bureau of Labor Statistics, *Second Annual Report*, 1871, p. 612;
Ohio Bureau of Labor Statistics, *First Annual Report*, 1877, pp. 197–198; Maryland
Bureau of Industrial Statistics, *First Biennial Report*, 1884–1885, p. 45; Massachusetts
Unemployed Board, *Report*, Part IV, 1895, pp. v–vii, xl, xlv.
38.   United States, Department of Commerce and Labor, *Eleventh Special Report
of the Commissioner of Labor, Regulation and Restriction of Output* (Washington,
D. C., 1904); Stanley B. Mathewson, *Restriction of Output Among Unorganized
Workers* (New York, 1931); Sumner Slichter, *Union Policies and Industrial Manage-
ment* (Washington, D. C., 1941), chs. 6, 7; Lloyd Ulman, *The Rise of the National
Trade Union* (Cambridge, 1955), pp. 526–533 and ch. 17; William Haber, *Industrial
Relations in the Building Industry* (Cambridge, 1930), ch. 8; Jesse Robinson, *The*

*Amalgamated Association of Iron, Steel and Tin Workers* (Baltimore, 1920), chs. 7, 8; David P. Smelser, *Unemployment and American Trade Unions* (Baltimore, 1919), pp. 46-53.

39.     Address to the Coopers of North America, Robert Schilling Papers, Wisconsin State Historical Society, Box 1. Also see Cornelius Keefe, President of the Tin and Sheet Iron Workers Association in *Fincher's Trades' Review*, May 1864, p. 102; *The Labor Holiday*, I (September 7, 1885), 2, as found in the A. R. Parsons Papers, Wisconsin State Historical Society, Box 2; *The Journal* (Metal Polishers, Buffers, Platers and Brass Workers' Union of North America), VI (January 1898), 441.

40.     *Fincher's Trades' Review*, December 10, 1864, p. 6.

41.     On the campaign for immigration restriction in the 1880's, see John Higham, *Strangers in the Land: Patterns of American Nativism, 1860-1925* (New Brunswick, N. J., 1955).

42.     New York Bureau of Labor Statistics, *Eighth Annual Report*, 1890, p. 698. Also see the interview with A. R. Parsons, *Canton Daily Repository*, February 6, 1886, as found in the Parsons Papers, Box 1.

43.     Originally published in *Labor World*. Found in *The Foremen's Advance Advocate*, V (1896), 503.

44.     New York Bureau of Labor Statistics, *Sixteenth Annual Report*, 1898, p. 1044.

45.     Eugene V. Debs, *Locomotive Firemen's Magazine*, XI (February 1887), 69-70; *The Painter*, II (July 1888), 2; H. H. Brown in *The Nationalist* (circa 1890) as found in John Samuel Papers, Wisconsin State Historical Society, Box 5; New York Bureau of Labor Statistics, *Eighth Annual Report*, 1890, p. 561.

46.     See Chapter IV.

47.     New York Bureau of Labor Statistics, *Eighth Annual Report*, 1890, p. 682; James O'Connell, President, International Association of Machinists, in *The Annals, American Academy of Political and Social Science*, XXVII (1906), 491-493.

48.     George Tracy, *History of the Typographical Union* (Indianapolis, 1913), pp. 1129-1144.

49.     *Journal of United Hatters of North America*, II (May 1900), 4.

50.     New York Bureau of Labor Statistics, *Eighth Annual Report*, 1890, p. 698; Mayor John Chase of Haverhill, Massachusetts in *International Wood-Worker*, VIII (December 1899), 135; Henry White, "Machinery and Labor," *The Annals, American Academy of Political and Social Science*, XX (1902), 229-230.

51.     See Sidney Fine, *Laissez-Faire and the General-Welfare State: A Study of Conflict in American Thought, 1865-1901* (Ann Arbor, 1956) and Henry Commager, *The American Mind: An Interpretation of American Thought and Character Since the 1880's* (New Haven, 1952).

52.     Herbert Baker, single taxer and President of the Connecticut Land and Labor League, in Connecticut Bureau of Labor Statistics, *Third Annual Report*, 1887, p. 315.

53.     On the campaigns for unemployment insurance in the twentieth century, see David Nelson, *Unemployment Insurance: The American Experience, 1915-1935* (Madison, Wis., 1969), and Irwin Yellowitz, "The Origins of Unemployment Reform in the United States," *Labor History*, IX (Fall 1968), 338-360. On relief activities for the unemployed, see Leah Feder, *Unemployment Relief in Periods of Depression: A Study of Measures Adopted in Certain American Cities, 1857 Through 1922* (New York, 1936).

54.     Quotations from John Gregory, "The Problem of the Unemployed," *The Independent*, XXXIX (November 10, 1887), 1443. Also see Massachusetts Bureau of Labor Statistics, *Twenty-Fourth Annual Report*, 1893, pp. 250-251; Massachusetts Unemployed Board, *Report*, Part IV, p. iii; B. O. Flower, "Emergency Measures Which Would Have Maintained Self-Respecting Manhood," *Arena*, IX (April 1894), 822-826; Washington Gladden, "What to Do with the Workless Man," National Conference of Charities and Correction, *Proceedings*, 1899, pp. 141-152; John Commons, "The Right to Work," *Arena*, XXI (February 1899), 131-142. For the

traditional view, see Davis R. Dewey, "Irregularity of Employment," *Publications of the American Economic Association,* IX (1894), 53–67; James Mavor, "Labor Colonies and the Unemployed," *Journal of Political Economy,* II (1893–1894), 26–53; D. M. Means, "The Dangerous Absurdity of State Aid," *Forum* (May 1894), 287–296.

55. [George Gunton], "Does Invention Lessen Employment?" *Gunton's Magazine,* XXIV (May 1898), 331.

56. Commons, "Right to Work," p. 140.

57. Charles Tuttle, "The Workman's Position in the Light of Economic Progress," American Economic Association, *Papers and Proceedings of the Fourteenth Annual Meeting, 1901* (1902), pp. 207–211.

58. Ibid., p. 230.

59. Ibid., p. 223.

60. Ibid., p. 233.

61. William F. Willoughby, *Workingmen's Insurance* (New York, 1898), pp. 375–376.

62. The labor movement of the 1860's had been much more open to political action as a basic weapon. See David Montgomery, *Beyond Equality: Labor and the Radical Republicans, 1862–1872* (New York, 1967).

63. Louis Reed, *The Labor Philosophy of Samuel Gompers* (New York, 1930); Fred Greenbaum, "The Social Ideas of Samuel Gompers," *Labor History,* VII (Winter 1966), 35–61; Roy Lubove, *The Struggle for Social Security, 1900–1935* (Cambridge, 1968), ch. 1.


CHAPTER III


1. William Sullivan, *The Industrial Worker in Pennsylvania, 1800–1840* (Harrisburg, 1955), pp. 29–83; Norman Ware, *The Industrial Worker, 1840–1860* (Boston, 1924), pp. 26–70; Walter Hugins, *Jacksonian Democracy and the Working Class: A Study of the New York Workingmen's Movement, 1829–1837* (Stanford, Calif., 1960), pp. 53–56; David Montgomery, "The Working Classes of the Pre-Industrial American City, 1780–1830," *Labor History,* IX (Winter 1968), 5–9; Raymond Mohl, *Poverty in New York, 1783–1825* (New York, 1971); Edward Pessen, *Most Uncommon Jacksonians: The Radical Leaders of the Early Labor Movement* (Albany, 1967), pp. 3–8. Also see Pessen, "The Egalitarian Myth and the American Social Reality: Wealth, Mobility and Equality in the 'Era of the Common Man,'" *American Historical Review,* LXXVI (October 1971), 989–1034.

2. On the traditional success ethic, see Irvin Wyllie, *The Self-Made Man in America: The Myth of Rags to Riches* (New Brunswick, N. J., 1954) and John Cawelti, *Apostles of The Self-Made Man* (Chicago, 1965). For an interesting case of mobility from skilled labor to entrepreneur, see Herbert Gutman, "The Reality of the Rags-to-Riches 'Myth': The Case of Patterson, New Jersey Locomotive, Iron and Machinery Manufacturers, 1830–1880" in Stephan Thernstrom and Richard Sennett (eds.), *Nineteenth-Century Cities: Essays in the New Urban History* (New Haven, 1969), pp. 98–124. Paul Faler's "Workingmen, Mechanics and Social Change: Lynn, Massachusetts, 1800–1860," Ph.D. dissertation, University of Wisconsin, 1971, examines the conversion of skilled journeymen artisans into employees at work in factories.

3. *The Toiler,* May 23, 1874, p. 1.

4. Maryland Bureau of Industrial Statistics, *First Biennial Report,* 1884–1885, p. 11.

5. Henry George, *Social Problems* (Chicago, 1883), pp. 29–42. Quotation from pp. 37–38.

6. Richard Ely, *The Labor Movement in America* (New York, 1886), pp. 92–95.

7. Peter J. McGuire, in United States Senate, Committee on Education and Labor,

*Report upon the Relations Between Labor and Capital,* 4 vols. (Washington, D. C., 1885), I, p. 356. Also see the *National Labor Tribune,* January 27, 1877, p. 2.

8.    *Journal of United Labor,* I (November 1880), 69.

9.    Commissioner John Lamb, Minnesota Bureau of Labor Statistics, *First Biennial Report,* 1887–1888, pp. 230–231.

10.    On producers' cooperation, see *Johns Hopkins University Studies in Historical and Political Science,* VI (1888); G. O. Virtue, "The Co-operative Coopers of Minneapolis," *Quarterly Journal of Economics,* XIX (August 1905), 527–544; Gerald Grob, *Workers and Utopia: A Study of Ideological Conflict in the American Labor Movement, 1865–1900* (Evanston, Ill., 1961), pp. 19–21, 43–48; Norman Ware, *The Labor Movement in the United States, 1860–1890* (New York, 1929), pp. 320–333; John Commons et al., *History of Labor in the United States,* II (New York, 1918), pp. 53–56, 110–112, 171–175.

11.    Michigan Bureau of Labor and Industrial Statistics, *First Annual Report,* 1883, p. 197; Iowa Bureau of Labor Statistics, *First Biennial Report,* 1884–1885, p. 210; *Second Biennial Report,* 1886–1887, p. 200; *Third Biennial Report,* 1888–1889, p. 97; Kansas Bureau of Labor and Industrial Statistics, *Fourth Annual Report,* 1888, pp. 130–131; Thomas McGuire, United States Senate, *Report on Labor and Capital,* I, p. 771.

12.    Kansas Bureau of Labor and Industrial Statistics, *Fifth Annual Report,* 1889, p. 67.

13.    *Labor Standard,* May 12, 1877, p. 3; Maryland Bureau of Industrial Statistics, *First Biennial Report,* 1884–1885, p. 10.

14.    Boot and Shoe Workers' Union, *Proceedings,* 1896, p. 15.

15.    For Gompers's views on large corporations and trusts, see Grob, *Workers and Utopia,* pp. 183–184; Louis S. Reed, *The Labor Philosophy of Samuel Gompers* (New York, 1930), p. 18; Bernard Mandel, *Samuel Gompers* (Yellow Springs, Ohio, 1963), pp. 137–139; Fred Greenbaum, "The Social Ideas of Samuel Gompers," *Labor History,* VII (Winter 1966), p. 48. Also see Louis Galombos, "A. F. L.'s Concept of Big Business: A Quantitative Study of Attitudes toward the Large Corporation, 1894–1931," *Journal of American History,* LVII (March 1971), 847–863.

16.    Edward Young, *Labor in Europe and America* (Philadelphia, 1875), p. 178.

17.    Pennsylvania Bureau of Statistics, *Second Annual Report,* 1873–1874, p. 428.

18.    United States Senate, *Report on Labor and Capital,* I, p. 758.

19.    Edward T. Peters, "Some Economic and Social Effects of Machinery," American Association for the Advancement of Science, *Proceedings,* 1884, pp. 639–640.

20.    See Herbert Gutman, "Labor's Response to Modern Industrialism," in Howard Quint et al., *Main Problems in American History,* II (Homewood, Ill., 1964), pp. 75–78.

21.    On the changing definition of success, see Stephan Thernstrom, *Poverty and Progress: Social Mobility in a Nineteenth-Century City* (Cambridge, 1964). Also see Stuart Blumin, "Mobility and Change in Ante-Bellum Philadelphia" in Thernstrom and Sennett (eds.), *Nineteenth-Century Cities,* pp. 165–208.

22.    Testimony of Horace Binney Sargent, United States House of Representatives, *Investigation by a Select Committee of the House of Representatives Relative to the Causes of the General Depression in Labor and Business; And as to Chinese Immigration* (Washington, D. C., 1879), p. 413.

23.    Michigan Bureau of Labor and Industrial Statistics, *First Annual Report,* 1883, pp. 181–182.

24.    Terence V. Powderly, *Thirty Years of Labor* (Columbus, Ohio, 1889), p. 127.

25.    Ibid., p. 156. The same argument was offered by John Butler of the Knights of Labor in Pennsylvania, Secretary of Internal Affairs, *Fifteenth Annual Report,* 1887, Part III (Industrial Statistics), p. G29.

26.    Ohio Bureau of Labor Statistics, *First Annual Report,* 1877, p. 228.

27.    George E. McNeill, "The Struggle for Life," *Work and Wages,* I (November 1886), 3.

28.   Ohio Bureau of Labor Statistics, *Second Annual Report*, 1878, p. 176.
29.   United Association of Journeymen Plumbers, Gas Fitters, Steam Fitters and Steam Fitters' Helpers, *Proceedings*, 1897, pp. 79–81. On the opposition of the plasterers to the use of precast cornices, see United States, Department of Commerce and Labor, *Eleventh Special Report of the Commissioner of Labor, Regulation and Restriction of Output* (Washington, D. C., 1904), pp. 302–303.
30.   See the *Painters' Journal*, VI (September 1892), 8. For other examples of this argument, see the remarks of a horseshoer in the *Labor Standard*, February 3, 1877, p. 1; the remarks of a currier in Massachusetts Bureau of Labor Statistics, *Tenth Annual Report*, 1879, p. 133; the boycott notice of the Can Makers' Mutual Protective Association of Baltimore, 1885, against the use of machine-made cans in John Samuel Papers, Wisconsin State Historical Society, Box 1; the discussion at the convention of the New England Lasters' Protective Union in its *Proceedings*, April 1889, p. 43; and the comments in the *Stone Cutters' Journal*, XVI (November 1902), 7.
31.   *Painters' Journal*, IV (August–September 1890), 7. Also see Maryland Bureau of Industrial Statistics, *First Biennial Report*, 1884–1885, p. 48.
32.   *The Garment Worker*, II (December 1895), 6; III (January 1896), 10; IV (January 1898), 5.
33.   *Journal of United Hatters of North America*, I (August 1898), 1, 4.


## CHAPTER IV

1.   *Fincher's Trades' Review*, August 15, 1863, p. 43.
2.   Ibid., May 6, 1865, p. 91; August 5, 1865, p. 76.
3.   A. D. Richardson, "Making Watches by Machinery," *Harper's New Monthly Magazine*, XXXIX (July 1869), 182.
4.   Ira Steward, "Meaning of the Eight Hour Movement," Ira Steward Papers, Wisconsin State Historical Society, Box 3.
5.   Untitled essay, Steward Papers, Box 3, p. 10. Also see George McNeill's strong support of Steward's theory in Massachusetts Bureau of Labor Statistics, *Second Annual Report*, 1871, pp. 559–564.
6.   "Unemployment," Steward Papers, Box 3, p. 2.
7.   *Labor Standard*, April 14, 1877, p. 2.
8.   Connecticut Bureau of Labor Statistics, *First Annual Report*, 1874, p. 51; *National Labor Tribune*, December 4, 1875, p. 1. The Furniture Workers' Union of Chicago stated that "Our first aim must be to restore a reasonable proportion between production and consumption, and as we cannot directly augment the ability to consume, we must therefore endeavor to reduce the production." *The Socialist*, May 20, 1876, p. 1.
9.   *National Labor Tribune*, September 2, 1876, p. 1.
10.   *The Socialist*, April 15, 1876, p. 2; *National Labor Tribune*, January 27, 1877, p. 2.
11.   *Machinists' and Blacksmiths' International Journal*, VIII (December 1870), 47; *Workingman's Advocate*, January 11, 1868, p. 2; Massachusetts Bureau of Labor Statistics, *Second Annual Report*, 1871, pp. 573–574.
12.   *Fincher's Trades' Review*, June 4, 1864, p. 2.
13.   Ibid., July 25, 1863, p. 31.
14.   United States, Department of Commerce and Labor, *Eleventh Special Report of the Commissioner of Labor, Regulation and Restriction of Output* (Washington, D. C., 1904), p. 491; Harold Aurand, *From the Molly Maguires to the United Mine Workers: The Social Ecology of an Industrial Union, 1869–1897* (Philadelphia, 1971), pp. 15–16; Clifton Yearley Jr., *Enterprise and Anthracite: Economics and Democracy in Schuylkill County, 1820–1875* (Baltimore, 1961), pp. 153–156.

15. Yearley, *Enterprise and Anthracite*, pp. 151-153, 158-164. The anthracite area of Pennsylvania was divided into three general regions: northern (Wyoming) of 176 square miles; middle (Lehigh) comprising two fields, together equaling 133 square miles; and southern (Schuylkill) covering 149 square miles. See Aurand, *Molly Maguires to U. M. W.*, pp. 5-8.

16. Edward Wieck, *The American Miners' Association: A Record of the Origins of Coal Miners' Unions in the United States* (New York, 1940), p. 75.

17. Aurand, *Molly Maguires to U. M. W.*, p. 66.

18. Wieck, *American Miners Association*, pp. 97-101.

19. Although restriction of output in the American anthracite fields during the 1860's was led by the miners' trade union, unorganized workers may also restrain production effectively through informal, but well understood, limits. On this point, see Stanley B. Mathewson, *Restriction of Output among Unorganized Workers* (New York, 1931).

20. G. O. Virtue, "The Anthracite Mine Laborers," United States, Bureau of Labor, *Bulletin*, No. 13 (1897), pp. 731-732.

21. *Workingman's Advocate*, August 22, 1868, p. 2.

22. Joseph Patterson, "The Old W. B. A. Days," Historical Society of Schuylkill County, *Publications*, II (1910), p. 356. Aurand, *Molly Maguires to U. M. W.*, p. 69.

23. *Workingman's Advocate*, May 29, 1869, p. 3.

24. John Brady to the Convention of the National Labor Union, *Workingman's Advocate*, August 27, 1870, p. 4; Letter from John Siney, President of the W. B. A., ibid., February 25, 1871, p. 3. Yearley, *Enterprise and Anthracite*, p. 186.

25. As found in Andrew Roy, *A History of the Coal Miners of the United States: From the Development of the Mines to the Close of the Anthracite Strike of 1902* (Columbus, Ohio, 190[3]), pp. 80-81.

26. Hendrick Wright, *A Practical Treatise on Labor* (New York, 1871), pp. 136-140.

27. *Workingman's Advocate*, March 20, 1869, p. 4.

28. Aurand, *Molly Maguires to U. M. W.*, pp. 69-73, 79-80; Samuel Eliot, "Relief of Labor," *Journal of Social Science*, IV (1871), 147.

29. Yearley, *Enterprise and Anthracite*, pp. 186-192; Aurand, *Molly Maguires to U. M. W.*, pp. 71-77.

30. American Workman, March 11, 1871, p. 5 (typescript in Wisconsin State Historical Society, Labor Collection, 11A); "The True History of the Coal Trouble," *The Nation*, XII (March 9, 1871), 154.

31. Aurand, *Molly Maguires to U. M. W.*, pp. 77, 84-93; Yearley, *Enterprise and Anthracite*, pp. 202-209; Wayne Broehl, *The Molly Maguires* (Cambridge, 1964), pp. 109-110.

32. *National Labor Tribune*, March 18, 1876, p. 1.

33. See Davis's argument, ibid., April 14, 1877, p. 1, and a response by D. McLaughlin of Braidwood, Illinois, ibid., April 28, 1877, p. 2.

34. United States, Labor Commissioner, *Eleventh Special Report*, 1904, pp. 487-505. Also see United States House of Representatives, *Labor Troubles in the Anthracite Regions of Pennsylvania, 1887-1888* (Washington, D. C., 1889), pp. xlvi-liv; Arbitration between the Coal Operators and Miners in the Northern District of Illinois, Testimony, July 25, 1889, p. 32 (Wisconsin State Historical Society, Labor Collection, 14A, Box 2).

35. Chris Evans, *History of United Mine Workers of America* (Vol. 1, 1860-1890; Vol. 2, 1890-1900) [1918], I, pp. 140, 199. Also see Ohio Bureau of Labor Statistics, *Ninth Annual Report*, 1885, pp. 55-56.

36. For McBride's statement, see Evans, *United Mine Workers*, I, pp. 483-484. The action of the convention is found on p. 490.

37. See Chapter VI.

38. *Workingman's Advocate*, August 6, 1870, p. 1.

39.   *National Labor Tribune,* May 26, 1877, p. 1; *Labor Standard,* September 2, 1877, p. 2; September 21, 1878, p. 1; United States House of Representatives, *Investigation by a Select Committee of the House of Representatives Relative to the Causes of the General Depression in Labor and Business; And as to Chinese Immigration* (Washington, D. C., 1879), pp. 115–116, 249; Ohio Bureau of Labor Statistics, *Second Annual Report,* 1878, p. 243. The same argument was used to explain the depression of 1893. See President Samuel Gompers's report, American Federation of Labor, *Proceedings,* 1893, p. 11; Walter S. Logan, *An Argument for an Eight-Hour Law* (New York, 1894), pp. 18–19; New York Bureau of Labor Statistics, *Eleventh Annual Report,* 1893, p. 3191.

40.   *Labor Standard,* May 12, 1877, p. 3; United States House of Representatives, *Investigation of Depression,* 1879, p. 412.

41.   United States House of Representatives, *Investigation of Depression,* 1879, pp. 48–49.

42.   *Coopers' Journal,* III (August 1872), 482–483.

43.   United States Department of Labor, Bureau of Labor Statistics, *Bulletin,* No. 441 (Boris Stern, "Productivity of Labor in the Glass Industry"), July 1927, p. 110.

44.   Pearce Davis, *The Development of the American Glass Industry* (Cambridge), 1949), p. 157.

45.   American Flint Glass Workers' Union, *History [of] American Flint Glass Workers' Union of North America* (Toledo, 1957), p. 8. This volume includes a reprint of a history of the union prepared in 1910 by a union officer, Thomas W. Rowe.

46.   U. S. Bureau of Labor Statistics, *Bulletin,* No. 441, p. 110.

47.   *History of Flint Glass Workers,* p. 8.

48.   *Iron Age,* September 20, 1877, p. 3; United States, Labor Commissioner, *Eleventh Special Report,* 1904, p. 643.

49.   President William J. Smith of the American Flint Glass Workers' Union in Pennsylvania, Secretary of Internal Affairs, *Sixteenth Annual Report,* 1888, Part III (Industrial Statistics), pp. 5–7.

50.   *National Labor Tribune,* December 7, 1878, p. 1.

51.   *Iron Age,* February 20, 1879, p. 11.

52.   For terms of the settlement, see ibid., July 3, 1879, p. 26; *National Labor Tribune,* July 5, 1879, p. 4; Pennsylvania, Secretary of Internal Affairs, *Sixteenth Annual Report,* 1888, Part III (Industrial Statistics), p. 7.

53.   *John Swinton's Paper,* March 23, 1884, p. 1. Also see Davis, *Glass Industry,* pp. 128–130.

54.   Andrew C. Robertson, who was a member of the Druggists' Ware Glass Blowers' League, in United States Senate, *Investigation of Labor Troubles* (Washington, D. C., 1893), p. 227.

55.   American Flint Glass Workers' Union, *Proceedings,* 1896, pp. 174–177; *Proceedings,* 1897, p. 37; United Green Glass Workers' Association, *Proceedings,* 1895, pp. 24–28, 74–78.

56.   United States, Labor Commissioner, *Eleventh Special Report,* pp. 601–606.

57.   American Flint Glass Workers' Union, *Proceedings,* 1889, pp. 26–28.

58.   *Journal of United Labor,* February 19, 1887, p. 2292. Also see *John Swinton's Paper,* March 9, 1884, p. 2.

59.   Michigan Bureau of Labor and Industrial Statistics, *Second Annual Report,* 1884, p. 116. On the underconsumption argument, also see Amalgamated Association of Iron and Steel Workers of the United States, *Proceedings,* 1885, p. 1612; "The True Cause of Hard Times," *The Nation,* XL (February 26, 1885), 179; S. M. Jelley, *The Voice of Labor* (Philadelphia, 1888), p. 115; New York Bureau of Labor Statistics, *Eighth Annual Report,* 1890, pp. 701, 703; *Foremen's Advance Advocate,* III (June 1894), 418.

60.   Carroll Wright, "The Relation of Production to Productive Capacity," *Forum,* XXIV (February 1898), 670–674.

61.  *Justice* (New York), December 20, 1884, p. 9.
62.  For a clear statement of this position, see Hiram S. Maxim, "Automatic Machinery: The Secret of Cheap Production," *The Engineering Magazine,* XIV (January 1898), 593–595.

## CHAPTER V

1.  International Workingmen's Association Papers, Wisconsin State Historical Society, Box 2, Clippings [1872].
2.  For a discussion of cigar manufacture in the nineteenth century, see Willis Baer, *The Economic Development of the Cigar Industry in the United States* (Lancaster, Pa., 1933), pp. 80–89.
3.  *Workingman's Advocate,* May 27, 1871, p. 3.
4.  See John Commons et al., *History of Labor in the United States,* II (New York, 1918), pp. 71–74.
5.  *Workingman's Advocate,* June 3, 1871, p. 3. Also see August 19, 1871, p. 3. The *Workingman's Advocate* was the official organ of the Cigar Makers' International Union from 1869 to 1875, and it is the most complete source on the activities of the union in this period.
6.  Ibid., November 25–December 2, 1871, p. 1.
7.  Ibid., p. 4.
8.  Ibid., May 27, 1871, p. 2.
9.  See the union's constitution of 1870, article XI, section 4, ibid., November 5–12, 1870, p. 4. For approval of the constitution by a referendum of the locals, see ibid., December 31, 1870, pp. 1, 4.
10.  Commons, *History of Labor,* II, pp. 72–73.
11.  *Workingman's Advocate,* October 12, 1872, p. 1.
12.  See President Edwin Johnson's report in *Workingman's Advocate,* October 12, 1872, p. 1. Also April 15, 1871, p. 3.
13.  Ibid., November 25–December 2, 1871, p. 2.
14.  Ibid., October 12, 1872, pp. 1, 4.
15.  Ibid., September 13, 1873, p. 2. For approval of the constitution by a referendum of the locals, see December 13, 1873, p. 4.
16.  Ibid., September 25, 1875, p. 1.
17.  Ibid.
18.  Constitution, article IV, section 1, as found in the *Cigar Makers' Official Journal,* October 10, 1881, p. 8.
19.  United States, Department of Commerce and Labor, *Eleventh Special Report of the Commissioner of Labor, Regulation and Restriction of Output* (Washington, D. C., 1904), pp. 568–569.
20.  Sumner Slichter, *Union Policies and Industrial Management* (Washington, D. C., 1941), pp. 216–221.
21.  United States Commissioner of Labor, *Eleventh Special Report,* pp. 565–566.
22.  Ibid., p. 574.
23.  United States, Works Progress Administration, National Research Project on Reemployment Opportunities and Recent Changes in Industrial Techniques, *Report B-4* (W. D. Evans, "Effects of Mechanization in Cigar Manufacture"), Washington, D. C., May 1938.
24.  The local option policy was not abandoned until 1923.
25.  Franklin E. Coyne, *The Development of the Cooperage Industry in the United States, 1620–1940* (Chicago, 1940), pp. 18, 21.
26.  "Among the Nail-Makers," *Harper's New Monthly Magazine,* XXI (July 1860), 164.
27.  Ohio Bureau of Labor Statistics, *First Annual Report,* 1877, p. 222.

28.    Coyne, *Cooperage Industry*, pp. 23–24; G. O. Virtue, "The Co-operative Coopers of Minneapolis," *Quarterly Journal of Economics*, XIX (August 1905), 538.

29.    Albert Shaw, "Cooperation in a Western City," *Publications of the American Economic Association*, I (September 1886), 23–24.

30.    Missouri Bureau of Labor Statistics, *Fourth Annual Report*, 1882, p. 117.

31.    United States, Department of Labor, *Thirteenth Annual Report of the Commissioner of Labor*, 2 vols. (Washington, D. C., 1898), II, pp. 933–941.

32.    Virtue, "Cooperative Coopers," p. 543.

33.    Coyne, *Cooperage Industry*, pp. 18, 22.

34.    Commons, *History of Labor*, II, pp. 74–75.

35.    *Coopers' Journal*, II (May 1871), 200–201; III (February 1872), 98. Also see *National Labor Tribune*, October 9, 1875, p. 1.

36.    *Coopers' Journal*, III (March 1872), 144; (October 1872), 622.

37.    Ibid., III (April 1872), 214.

38.    Ibid., II (January 1871), 73; (February 1871), 102; (November 1871), 404–406.

39.    On the shorter day, see Chapter IX.

40.    Coopers' International Union, *Constitution*, 1897, p. 24; *Coopers' International Journal*, X (December 1900), 10, 12; (March 1901), 5.

41.    Coopers' International Union, *Constitution*, 1899, pp. 33–34.

42.    Ibid., 1902, p. 11; 1904, p. 18.

43.    Coopers' International Union, *Proceedings*, 1902 as found in *Coopers' International Journal*, XI (October 1902), 428, 440; *Proceedings*, 1904 in XIII (September 1904), 478.

## CHAPTER VI

1.    For an excellent discussion of the response of the stonecutters' unions, see George Barnett, *Chapters on Machinery and Labor* (Cambridge, 1926), ch. 2.

2.    For the specific rules, see ibid., pp. 37–38, 47–49.

3.    On the efforts to control the shipment of machine-planed stone, see ibid., pp. 39–46.

4.    Ibid., pp. 52–59.

5.    The decision by the Amalgamated Meat Cutters and Butcher Workmen to take sides in a similar clash between butchers employed in western plants, with a more advanced technology, and those in Buffalo, New York, who sought to block innovation, nearly split the union. See David Brody, *The Butcher Workmen: A Study of Unionization* (Cambridge, 1964), p. 66.

6.    Barnett, *Machinery and Labor*, p. 46.

7.    On the planermen and the use of skilled union members as machine operators, see ibid., pp. 49–52.

8.    *Stone Cutters' Journal*, XI (July 1899), 11.

9.    Ibid., XVII (January 1903, Supplement), 20.

10.    Ibid., XVI (November 1902), 7.

11.    Ibid., XV (December 1901), 14; XI (November 1899), 12.

12.    Ibid., VIII (May 1896), 5.

13.    Ibid., XV (December 1901), 15.

14.    Ibid., XVI (April 1902), 15; also (September 1902), 17 and (December 1902), 15.

15.    Barnett, *Machinery and Labor*, p. 52.

16.    *Stone Cutters' Journal*, IX (October 1897), 9.

17.    Reprint of an article by Henry Gorse in *Stone*, ibid., XI (March 1899), 9. Also see XV (May 1901), 11.

18.    Ibid., XVI (March 1902), 15. In 1899, the reformer, Graham Taylor, claimed that the smaller employers in Chicago had joined the stonecutters in opposing the

planer. See United States Industrial Commission, *Report*, 19 vols. (Washington, D. C., 1900-1902), VIII, p. 542.
19.  *Stone Cutters' Journal*, XI (June 1899), 2.
20.  See Sumner Slichter, *Union Policies and Industrial Management* (Washington, D. C., 1940), chs. 7-9.
21.  On the printers, see Barnett, *Machinery and Labor*, ch. 1. On the glassblowers, see ibid., chs. 3, 4.
22.  Margaret L. Stecker, "The Founders, the Molders and the Molding Machine," *Quarterly Journal of Economics*, XXXII (February 1918), 278-308.
23.  United States, Labor Commissioner, *Eleventh Special Report*, 1904, p. 36.
24.  See Chapter V.
25.  Barnett, *Machinery and Labor*, pp. 72-79.
26.  For excellent discussions of the factors that motivated businessmen to innovate, see W. Paul Strassmann, *Risk and Technological Innovation: American Manufacturing During the Nineteenth Century* (Ithaca, N. Y., 1956) and Jacob Schmookler, *Invention and Economic Growth* (Cambridge, 1966). Also useful are W. E. G. Salter, *Productivity and Technical Change* (London, 1966) and Harry Jerome, *Mechanization in Industry* (New York, 1934).
27.  Arthur H. Cole and Harold F. Williamson, *The American Carpet Manufacture* (Cambridge, 1941), pp. 60-61.
28.  William Panschar, *Baking in America: Economic Development* (Evanston, Ill., 1956), pp. 68-69.
29.  S. McKee Rosen and Laura Rosen, *Technology and Society: The Influence of Machines in the United States* (New York, 1941), pp. 211-212.
30.  United States, National Resources Committee, *Technological Trends and National Policy* (Bernhard Stern, "Resistance to the Adoption of Technological Innovations," Washington, D. C., 1937), pp. 53-54.
31.  Rosen, *Technology and Society*, p. 221.
32.  American Flint Glass Workers' Union, *Proceedings*, 1898, pp. 50, 53; *Proceedings*, 1899, pp. 209, 214.
33.  Ibid., 1898, p. 54; 1899, pp. 32-33.
34.  Ibid., 1898, p. 54.
35.  Ibid., pp. 54-55.
36.  See Chapter V.
37.  American Flint Glass Workers' Union, *Proceedings*, 1898, pp. 26, 55.
38.  Glass Bottle Blowers' Association of the United States and Canada, *Proceedings*, 1903, p. 19. Also see *Proceedings*, 1900, pp. 49, 53.
39.  American Flint Glass Workers' Union, *Proceedings*, 1898, p. 23. Also see Glass Bottle Blowers' Association of the United States and Canada, *Proceedings*, 1902, p. 15.
40.  Mark Perlman, *The Machinists: A New Study in American Trade Unionism* (Cambridge, 1961), p. 24. On the general attitude of trade unions toward piece rates, see David McCabe, *The Standard Rate in American Trade Unions* (Baltimore, 1912).
41.  H. M. Norris, "Actual Experience with the Premium Plans," *The Engineering Magazine*, XVIII (January 1900), 583. Norris advocated a premium plan that he believed could increase productivity without raising some of the traditional opposition from workers to piece rates. For another discussion of this plan, see F. A. Halsey, "Experience with the Premium Plan for Paying for Labor," *American Machinist*, XXII (March 9, 1899), 17-20.
42.  "New Shop Methods a Corollary of Modern Machinery," *The Engineering Magazine*, XIX (June 1900), 370-372.
43.  International Association of Machinists, *Proceedings*, 1899 as found in the *Machinists' Monthly Journal*, XI (June 1899), 355-356. Also *Proceedings*, 1901 in XIII (July 1901), 465-466.
44.  Perlman, *Machinists*, p. 30. The vote was 1,610 in favor to 8,693 opposed.
45.  Gladys L. Palmer, *Union Tactics and Economic Change* (Philadelphia, 1932), pp. 91-99.

46. Ibid., p. 63.

47. Edward Wieck, *The American Miners' Association: A Record of the Origin of Coal Miners' Unions in the United States* (New York, 1940), p. 48.

48. Illinois, Bureau of Labor Statistics, *Fourteenth Annual Report*, 1895 (Statistics of Coal in Illinois: Supplemental Report), pp. xxxi–xxxii; United Mine Workers of America, *Proceedings*, 1908, p. 36.

49. Illinois Bureau of Labor Statistics, *Fifth Biennial Report*, 1888, pp. 339–340.

50. Illinois Bureau of Labor Statistics, *Sixth Biennial Report*, 1890, p. lxv.

51. The number of coal miners in the United States increased from 200,000 in 1900 to 600,000 in 1920. See Morton Baratz, *The Union and the Coal Industry* (New Haven, 1955), pp. 43–45.

52. Quotation from Illinois Bureau of Labor Statistics, *Third Biennial Report*, 1884, p. 576. Also see p. 434 and *Second Biennial Report*, 1882, pp. 36, 58, 81, 99.

53. On support from the community for strikes by coal miners, see Herbert Gutman, "The Worker's Search for Power: Labor in the Gilded Age," in Wayne H. Morgan (ed.), *The Gilded Age: A Reappraisal* (Syracuse, 1963), pp. 38–68.

54. National Federation of Miners and Mine Laborers of the United States and Territories, *Proceedings*, 1886, n. p.

55. Chris Evans, *History of United Mine Workers of America* [1918], II (1890–1900), p. 90.

56. Isadore Lubin, *Miners' Wages and the Cost of Coal* (New York, 1924), p. 186.

57. Evans, *History of United Mine Workers*, II, p. 166.

58. United Mine Workers of America, *Proceedings*, 1900, p. 24.

59. Ibid., 1901, pp. 46, 89, 107; 1902, pp. 45–46, 102; 1904, p. 32.

60. On this point, see Baratz, *Union and Coal Industry*, p. 53.

61. United Mine Workers of America, *Proceedings*, 1899, pp. 8–13; Joint Conference of Miners and Operators, Interstate Convention, *Proceedings*, 1899, p. 33; 1906, p. 144; 1910, p. 16; 1912, p. 15.

62. Joint Conference of Miners and Operators, *Proceedings*, 1899, p. 14; 1900, p. 75.

63. Ibid., 1901, p. 67.

64. Ibid., 1900, p. 22.

65. Lubin, *Miners' Wages*, pp. 107–108.

66. Ibid., pp. 109, 261–262.

67. My calculations of the differential are based upon data found in the United States Coal Commission, *Report*, Part V (Washington, D. C., 1925), pp. 183–185.

68. Baratz, *Union and Coal Industry*, pp. 41–42.

69. See David Brody, *Steelworkers in America: The Non-Union Era* (Cambridge, 1960), pp. 1–49.

70. United States House of Representatives, *Labor Troubles at Homestead, Pennsylvania* (Washington, D. C., 1893), p. 74.

71. Jesse Robinson, *The Amalgamated Association of Iron, Steel and Tin Workers* (Baltimore, 1920), p. 128.

72. Ibid., pp. 114–123.

73. Brody, *Steelworkers*, p. 58. Also see Carroll Wright, "The Amalgamated Association of Iron and Steel Workers," *Quarterly Journal of Economics*, VII (July 1893), 432.

74. United States House of Representatives, *Labor Troubles at Homestead*. For Frick's position, see pp. 19–21, 29, 163; for the Union's response, see pp. 186–187.

75. Edward Bemis, "The Homestead Strike," *Journal of Political Economy*, II (June 1894), 377.

76. For an account of the strike, see Leon Wolff, *Lockout: The Story of the Homestead Strike of 1892* (New York, 1965).

77. Robert Ozanne, *A Century of Labor-Management Relations at McCormick and International Harvester* (Madison, Wis., 1967), pp. 26–28.

78.  Cole and Williamson, *Carpet Manufacture*, pp. 57–58.
79.  Martha Taber, *A History of the Cutlery Industry in the Connecticut Valley* (Smith College Studies in History, XXXXI, 1955), pp. 113–114.
80.  *Boot and Shoe Recorder*, XVII (September 3, 1890), 87; XVIII (March 18, 1891), 95; XIX (July 8, 1891), 71; (August 19, 1891), 85; (September 16, 1891), 81; Lynn Lasters' Union, Papers, Baker Library, Harvard University, Minutes of the Board, I (March 23, 1898), 62; (March 25, 1898), 65; (April 6, 1898), 68–69; (April 13, 1898), 71; (June 28, 1899), 217; (July 6, 1899), 218; (July 12–19, 1899), 219–223; (August 16, 1899), 228.
81.  *Locomotive Firemen's Magazine*, XIV (May 1890), 388.
82.  Ohio Bureau of Labor Statistics, *Thirteenth Annual Report*, 1889, p. 66.
83.  *Foremen's Advance Advocate*, I (November 1892), 420.
84.  Maryland Bureau of Industrial Statistics, *First Annual Report*, 1892, p. 180; Brotherhood of Carpenters and Joiners of America, *Constitution*, 1886, p. 2.
85.  George Mowry, *The Era of Theodore Roosevelt, 1900–1912* (New York, 1958), pp. 99–104.
86.  On the situation in the building trades, see William Haber, *Industrial Relations in the Building Industry* (Cambridge, 1930).

## CHAPTER VII

1.  On the labor movement of the 1850's, see Carl Degler, "Labor in the Economy and Politics of New York City, 1850–1860: A Study of the Impact of Early Industrialism," Ph.D. dissertation, Columbia University, 1952. On apprenticeship, see James Motley, *Apprenticeship in American Trade Unions* (Baltimore, 1907); Lloyd Ulman, *The Rise of the National Trade Union* (Cambridge, 1955), pp. 305–321. On the building trades, see William Haber, *Industrial Relations in the Building Industry* (Cambridge, 1930).
2.  *Fincher's Trades' Review*, September 24, 1864, p. 66; *Coopers' Journal*, II (May 1871), 220; *Machinists' and Blacksmiths' International Journal*, IX (March 1872), 552–553; *Journal of United Hatters of North America*, I (November 1898), 4.
3.  *Fincher's Trades' Review*, May 7, 1864, p. 90.
4.  Ibid., August 13, 1864, p. 42; *Coopers' Journal*, II (May 1871), 220; *Workingman's Advocate*, June 26, 1869, p. 3.
5.  *Machinists' and Blacksmiths' International Journal*, IX (February 1872), 521; Iron Molders' International Journal (June 1874), 401 as found in a typescript, Wisconsin State Historical Society, Labor Collection, 14A, Box 1; The Printers' Circular (February 1871), 503–508, ibid., Box 2 (typescript). Also see Jonathan Grossman, *William Sylvis, Pioneer of American Labor: A Study of the Labor Movement During the Era of the Civil War* (New York, 1945), pp. 132–134.
6.  George Tracy, *History of the Typographical Union* (Indianapolis, 1913), pp. 119–123. Also see Bricklayers' National Union, *Proceedings*, 1868, p. 24.
7.  See the proceedings of the National Typographical Union. For examples of the apprenticeship regulations in other trades, during this period, see for the stonecutters, Journeymen Stone-Cutters' Association of the District of Columbia, *Constitution and By-Laws*, 1854, pp. 4, 6, and American Workman, January 20, 1872, p. 1 (typescript in Wisconsin State Historical Society, Labor Collection, 11A); for the cigar makers, Cigar Makers' Society of the State of Maryland, Minutes, September 7, 1860, p. 148; September 27, 1860, p. 153 (typescript, Johns Hopkins); for the ship caulkers, *Fincher's Trades' Review*, June 10, 1865, p. 13; for the bricklayers, *Workingman's Advocate*, August 1, 1868, p. 4; February 6, 1869, p. 1; March 26, 1870, p. 2; for the hatters, National Trade Association of Hat Finishers of the United States of America, Constitution, 1863, p. 9, as found in typescript, Wisconsin State Historical Society, Labor Collection, 14A.

8.   Charges of monopoly over the job even reached into labor and reform circles. See John Shedden, Minutes, Philadelphia Section, International Workingmen's Association, September 23, 1872, in International Workingmen's Association Papers, Wisconsin State Historical Society, Box 3; L. C. Hughes before the convention of the Machinists' and Blacksmiths' International Union in *Machinists' and Blacksmiths' International Journal*, VIII (December 1870), 45; President J. C. Whaley, National Labor Union, *Proceedings*, 1868, p. 7; Ezra Heywood at the New England Labor Reform Convention, *Workingman's Advocate*, May 27, 1871, p. 1.

9.   *Coopers' Journal*, IV (January 1873), 12. Also see Robert Schilling, "Address to the Coopers of North America" in Robert Schilling Papers, Wisconsin State Historical Society, Box 1.

10.  Letter from Hugh McLaughlin, December 27, 1870, in the *Vulcan Record*, I, No. 7 (1870), 31. Also see the reprint of the statement by President William Sylvis to the convention of the Iron Molders' International Union in *The International Journal*, I, n. s. (January 1867), 309.

11.  Massachusetts Bureau of Labor Statistics, *First Annual Report*, 1870, p. 320.

12.  *Fincher's Trades' Review*, July 18, 1863, p. 27.

13.  *Workingman's Advocate*, July 28, 1866, p. 4.

14.  Ibid., February 12, 1870, p. 1.

15.  Ibid., January 30, 1869, p. 2.

16.  Ira Steward, "Subdivision of Labor," Ira Steward Papers, Wisconsin State Historical Society, Box 3, p. 1. On the Crispins, see the valuable article by John P. Hall, "The Knights of St. Crispin in Massachusetts, 1869-1878," *Journal of Economic History*, XVIII (June 1958), 161-175. On the economic changes in the shoemaking industry, see Don E. Lescohier, *The Knights of St. Crispin: A Study in the Industrial Causes of Trade Unionism* (Madison, Wis., 1910), pp. 12-21.

17.  Massachusetts Bureau of Labor Statistics, *Second Annual Report*, 1871, pp. 604-616.

18.  American Workman (typescript), May 29, 1869, p. 6.

19.  Lescohier, *Knights of St. Crispin*, pp. 25-28; Hall, "Crispins in Massachusetts," p. 174.

20.  Massachusetts Bureau of Labor Statistics, *Second Annual Report*, 1871, p. 241.

21.  American Workman (typescript), April 23, 1870, p. 4.

22.  Ibid., May 22, 1869, p. 5.

23.  Ibid., June 19, 1869, p. 2.

24.  On the support for the Crispins by small manufacturers, see Massachusetts Bureau of Labor Statistics, *Eighth Annual Report*, 1877, pp. 21, 24.

25.  *The Socialist*, April 15, 1876, p. 2.

26.  Minnesota Bureau of Labor Statistics, *Fourth Biennial Report*, 1893-1894, pp. 228-229, 269-271.

27.  United States, Department of Commerce and Labor, *Eleventh Special Report of the Commissioner of Labor, Regulation and Restriction of Output* (Washington, D. C., 1904), pp. 669-672.

28.  Window Glass Workers, Local Assembly 300, Knights of Labor, *Proceedings*, 1889, pp. 11-17.

29.  Commissioner Thomas Bigham, Pennsylvania Bureau of Statistics, *Second Annual Report*, 1873-1874, p. 439.

30.  Robert A. Christie, *Empire in Wood: A History of the Carpenters' Union* (Ithaca, N. Y., 1956), pp. 5, 25-28, 79-82; Maryland Bureau of Industrial Statistics, *First Biennial Report*, 1884-1885, p. 45; *Third Annual Report*, 1894, p. 119; Missouri Bureau of Labor Statistics, *First Annual Report*, 1879, p. 54; Ohio Bureau of Labor Statistics, *Second Annual Report*, 1878, p. 176; *The Carpenter* (February 1899), 13.

31.  Maryland Bureau of Industrial Statistics, *Fifth Annual Report*, 1896, p. 39.

32.  *The Painter*, I (May 1887), 2; (August 1887), 2.

33.   Ibid., III (February 1889), 2.
34.   Frank Stockton, *The International Molders Union of North America* (Baltimore, 1921), pp. 170–179; Minnesota Bureau of Labor Statistics, *Fourth Biennial Report*, 1893–1894, pp. 246–248.
35.   Minnesota Bureau of Labor Statistics, *Fourth Biennial Report*, 1893–1894, p. 259.
36.   See *The Tailor*, III (August 1891), 2; (August 1893), 2; VIII (September 1897), 1. Journeymen Tailors' Union of America, *Constitution*, 1902, p. 11; Illinois Bureau of Labor Statistics, *Seventh Biennial Report*, 1892.
37.   Boycott notice in John Samuel Papers, Wisconsin State Historical Society, Box 12.
38.   Quotation from *Cincinnati Commercial*, May 23, 1874 as found in Herbert Gutman, "Social and Economic Structure and Depression: American Labor in 1873 and 1874," Ph.D. dissertation, University of Wisconsin, 1959, p. 137.
39.   Testimony before United States Senate, Committee on Education and Labor, *Report upon the Relations Between Labor and Capital*, 4 vols. (Washington, D. C., 1885), I, p. 593. On the impact of apprentices as cheap labor, see Illinois Bureau of Labor Statistics, *Second Biennial Report*, 1882, pp. 311ff.
40.   Missouri Bureau of Labor Statistics, *Second Annual Report*, 1880, p. 248. Also see Iowa Bureau of Labor Statistics, *First Biennial Report*, 1884–1885, pp. 204–207.
41.   District Assembly 149, Knights of Labor, *Proceedings*, 1887, pp. 6–16; New Jersey Bureau of Labor and Industrial Statistics, *Thirteenth Annual Report*, 1890, p. 419.
42.   United States Senate, *Report on Labor and Capital*, I, p. 668.
43.   United Association of Journeymen Plumbers, Gas Fitters, Steam Fitters and Steam Fitters' Helpers, *Proceedings*, 1897, pp. 71, 73.
44.   *Plumbers, Gas and Steam Fitters' Journal*, IV (October 1900), 15.
45.   New York Bureau of Labor Statistics, *Fourth Annual Report*, 1886, p. 219.
46.   Minnesota Bureau of Labor Statistics, *Fourth Biennial Report*, 1893–1894, pp. 132–137.
47.   Ibid., pp. 145–146.
48.   See Irwin Yellowitz, *Labor and the Progressive Movement in New York State, 1897–1916* (Ithaca, 1965).

## CHAPTER VIII

1.   There is little on the regulation of female and child labor in the nineteenth century. The early twentieth century has been studied. See John Commons et al., *History of Labor in the United States* (New York, 1935), III, pp. 399–500; Allen F. Davis, *Spearheads of Reform: The Social Settlements and the Progressive Movement, 1890–1914* (New York, 1967), ch. 7; Robert Bremner, *From the Depths: The Discovery of Poverty in the United States* (New York, 1956), pp. 76–80, 212–230; Jeremy Felt, *Hostages of Fortune: Child Labor Reform in New York State* (Syracuse, 1965); Walter Trattner, *Crusade for the Children: A History of the National Child Labor Committee and Child Labor Reform in America* (Chicago, 1970); Roger Walker, "The A. F. L. and Child-Labor Legislation: An Exercise in Frustration," *Labor History*, XI (Summer 1970), 323–340.
2.   On female and child labor in the early 1870's, see Herbert Gutman, "Social and Economic Structure and Depression: American Labor in 1873 and 1874," Ph.D. dissertation, University of Wisconsin, 1959, pp. 241–251. On child labor laws in Massachusetts, see Massachusetts Bureau of Labor Statistics, *First to Third Annual Reports*, 1870–1872. On New York, see Workingmen's Assembly of the State of New York, *Proceedings*, 1869–1871.

3.  Ohio Bureau of Labor Statistics, *Sixth Annual Report,* 1882, pp. 325–326, 329.
4.  National Typographical Union, Synopsis of Proceedings, 1854, p. 4 (typescript, Wisconsin State Historical Society). The vote was 17 to 9. On the relationship between the labor movement and the women's rights movement in the 1860's, see Israel Kugler, "The Women's Rights Movement and the National Labor Union, 1866–1872," Ph.D. dissertation, New York University, 1954. Also see Eleanor Flexner, *Century of Struggle: The Women's Rights Movement in the United States* (Cambridge, 1959), pp. 131–141; Jonathan Grossman, *William Sylvis, Pioneer of American Labor: A Study of the American Labor Movement During the Era of the Civil War* (New York, 1945), pp. 226–228.
5.  *Fincher's Trades' Review,* January 16, 1864, p. 28. Compare with the similar position in *The Laster,* March 15, 1892, p. 2.
6.  Ibid., May 21, 1864, p. 98.
7.  Ibid., October 1, 1864, p. 70; January 28, 1865, p. 34.
8.  National Labor Union, *Proceedings,* 1868, p. 7; Gerald Grob, *Workers and Utopia: A Study of Ideological Conflict in the American Labor Movement, 1865–1900* (Evanston, Ill., 1961), pp. 25–26.
9.  *Labor Standard,* July 14, 1878, p. 1.
10.  Knights of Labor, *Proceedings,* 1879, p. 103. Alse see Thomas Kidd, *Machine Wood-Worker,* I (November 1891), 5; Peter J. McGuire, December 7, 1891 as found in Edward Bemis Papers, Wisconsin State Historical Society, Box 2.
11.  *Labor Standard,* September 23, 1876, p. 2. Compare *Stone Cutters' Journal,* IX (October 1897), 10.
12.  United States House of Representatives, *Investigation by a Select Committee of the House of Representatives Relative to the Causes of the General Depression in Labor and Business; And as to Chinese Immigration* (Washington, D. C., 1879), pp. 430–432.
13.  New York Bureau of Labor Statistics, *Second Annual Report,* 1884, p. 331.
14.  *Locomotive Firemen's Magazine,* XI (September 1887), 522.
15.  Claus Andres, President of Local No. 88, in the *Cigar Makers' Official Journal,* XIII (November 1887), 6.
16.  On this point, compare the *National Labor Tribune,* January 27, 1877, p. 2, and William F. Willoughby, "Child Labor," *Publications of the American Economic Association,* V (March 1890), 176–179.
17.  Willoughby, "Child Labor," pp. 163–168.
18.  Massachusetts Bureau of Labor Statistics, *Sixth Annual Report,* 1875, p. 55.
19.  John Swinton's Paper, February 10, 1884, p. 4; Claire de Graffenreid, "Child Labor," *Publications of the American Economic Association,* V (March 1890), 261; Connecticut Bureau of Labor Statistics, *Third Annual Report,* 1887, p. 309.
20.  Massachusetts Bureau of Labor Statistics, *Sixth Annual Report,* 1875, p. 384.
21.  Neil J. Smelser, *Social Change in the Industrial Revolution: An Application of Theory to the British Cotton Industry* (Chicago, 1959), pp. 188–260.
22.  See above, Chapter III. For a fine statement of the dependency theme, see Jesse H. Jones, *Equity,* II (May 1875), 16.
23.  See Richard Sennett, *Families Against the City: Middle Class Homes of Industrial Chicago, 1872–1890* (Cambridge, 1970), chs. 5–7.
24.  Ibid., pp. 123–124.
25.  For examples of this line of argument, see Arthur T. Hadley, *Economics: An Account of the Relations Between Private Property and Public Welfare* (New York, 1896), pp. 345–346 and Dingman Versteeg, *Labor Saving Machinery and Progress* (New York, 1895), p. 34.
26.  Willoughby, "Child Labor," p. 187.
27.  Massachusetts Bureau of Labor Statistics, *Sixth Annual Report,* 1875, pp. 57–63. Also see A. J. Lathrop, *Labor Standard,* May 7, 1881, p. 6.
28.  Massachusetts Bureau of Labor Statistics, *Sixth Annual Report,* 1875, pp. 445–449

29.   United States Senate, Committee on Education and Labor, *Report upon the Relations Between Labor and Capital,* 4 vols. (Washington, D. C., 1885), II, p. 6.

30.   Connecticut Bureau of Labor Statistics, *First Annual Report,* 1874, p. 49.

31.   For examples of this approach, see the Federation of Organized Trades and Labor Unions of the United States, *Proceedings,* 1881, p. 18; Ohio Bureau of Labor Statistics, *Fifth Annual Report,* 1881, p. 89; *The Carpenter* (October 1889), 2.

32.   Massachusetts Bureau of Labor Statistics, *Sixth Annual Report,* 1875, pp. 67-111, 183-184. The quotation is from p. 183.

33.   We have noted some support for this policy in the labor movement during the 1860's, but it received less attention in later decades. However, Ben Butler stressed equal pay as his solution for the problem of female labor during his independent presidential campaign of 1884. See *Justice,* August 23, 1884, p. 10.

34.   American Federation of Labor, *Proceedings,* 1890, p. 40; 1891, p. 40.

35.   Ibid., 1894, pp. 31, 45.

36.   *Machine Wood-Worker,* III (January 1893), 13.

37.   Missouri Bureau of Labor Statistics, *Second Annual Report,* 1880, pp. 147, 151.

38.   New York Bureau of Labor Statistics, *Fourth Annual Report,* 1886, p. 24. Also see Commissioner A. D. Fassett, Ohio Bureau of Labor Statistics, *Thirteenth Annual Report,* 1889, p. 9.


## CHAPTER IX

1.   The Benevolent and Protective Association of the United Operative Mule Spinners of New England, Constitution and By-Laws, 1858, p. 1 (typescript, Wisconsin State Historical Society, Labor Collection, 14A).

2.   *Fincher's Trades' Review,* October 3, 1863, p. 70; Iron Molders' Journal, April 1874, pp. 321-323 (typescript, Wisconsin State Historical Society, Labor Collection, 14A, Box 1).

3.   See David Montgomery, *Beyond Equality: Labor and the Radical Republicans, 1862-1872* (New York, 1967), chs. 6, 8; John Commons et al., *History of Labor in the United States* (New York, 1918), II, ch. 4; Jonathan Grossman, *William Sylvis, Pioneer of American Labor: A Study of the American Labor Movement During the Era of the Civil War* (New York, 1945), pp. 129-132, 238-247. On the passage of the eight hour law in New York, see James C. Mohr, *The Radical Republicans and Reform in New York During Reconstruction* (Ithaca, N. Y., 1973), pp. 120-139.

4.   On the campaigns for eight hours in the 1880's, see Gerald Grob, *Workers and Utopia: A Study of Ideological Conflict in the American Labor Movement, 1865-1900* (Evanston, Ill., 1961), pp. 73-78, 149-151; Marion Cahill, *Shorter Hours: A Study of the Movement Since the Civil War* (New York, 1932), pp. 40-49, 152-164; Commons, *History of Labor,* II, pp. 375-386, 475-479; Philip Taft, *The A. F. of L. in the Time of Gompers* (New York, 1957), pp. 142-145; Sidney Fine, "The Eight-Hour Day Movement in the United States, 1888-1891," *Mississippi Valley Historical Review,* XL (December 1953), 441-462.

5.   American Federation of Labor, *Proceedings,* 1887, p. 9.

6.   New York Bureau of Labor Statistics, *Twelfth Annual Report,* 1894, p. 237.

7.   *Fincher's Trades' Review,* June 27, 1863, p. 14; February 20, 1864, p. 47; American Workman, July 31, 1869, p. 4 (typescript in Wisconsin State Historical Society, Labor Collection 11A); *Labor Standard,* November 2, 1878, p. 1; Speech by Thomas E. Hill, in A. R. Parsons, Papers, April 1885, Wisconsin State Historical Society, Box 2; American Federation of Labor, *Proceedings,* 1889, p. 29; *Machine Wood-Worker,* III (January 1893), 13.

8.   United States Senate, Committee on Education and Labor, *Report upon the Relations Between Labor and Capital,* 4 vols. (Washington, D. C., 1885) I, Layton, p. 37, Strasser, p. 459.

9.    George E. McNeill, *The Eight Hour Primer* (American Federation of Labor, Eight Hour Series, no. 1, Washington, D. C., 1888), p. 1. Also see the *National Labor Tribune*, December 2, 1876, p. 1, and New York Bureau of Labor Statistics, *Eighth Annual Report*, 1890, p. 539.

10.    Ohio Bureau of Labor Statistics, *Second Annual Report*, 1878, p. 276.

11.    *The Bricklayer and Mason*, I (April 20, 1899), 4. Also see the comments by A. T. Crane of the Crane Manufacturing Company, October 1885, in Parsons Papers, Box 1, and by Francis A. Walker, January 1890, in the Edward Bemis Papers, Wisconsin State Historical Society, Box 2.

12.    *Reports of Commissioners on the Hours of Labor* (Massachusetts House Document no. 44), Boston, 1867, pp. 24–25; Massachusetts Bureau of Labor Statistics, *Twelfth Annual Report*, 1881, pp. 384–394; *Milwaukee Sentinel*, May 2, 1886, p. 6.

13.    "The Eight-Hour Delusion," *The Nation*, III (November 22, 1866), 412–413; A. S. Cameron, *The Eight Hour Question* (New York, 1872), pp. 2, 8; "The Eight-hour Working-day," *The Century*, XXXIII (December 1886), 318.

14.    "Labor Strikes," *Harper's New Monthly Magazine*, XXXXVII (June 1873), 142.

15.    Letter to the editor from "A Journeyman Mechanic" of Elmira, New York, in the *New York Tribune*, as found in John Samuel Papers, Wisconsin State Historical Society, Box 10.

16.    American Federation of Labor, *Proceedings*, 1889, pp. 29–30.

17.    For the conservative argument, see *Massachusetts Report on Hours*, 1867, p. 28; Lyman Atwater, "The Labor Question in Its Economic and Christian Aspects," *Presbyterian Quarterly and Princeton Review*, n.s. I (1872), 480–481; Arthur T. Hadley, *Economics: An Account of the Relations Between Private Property and Public Welfare* (New York, 1896), pp. 405–406.

18.    McNeill, *Eight Hour Primer*, p. 15.

19.    Reprint of an article from the *New York Dispatch* in *The Carpenter* (July 1893), 4.

20.    On this point, see the testimony of T. B. Shannon, United States House of Representatives, *Investigation by a Select Committee of the House of Representatives Relative to the Causes of the General Depression in Labor and Business; And as to Chinese Immigration* (Washington, D. C., 1879), pp. 250–251; *The Garment Worker*, V (December 1898), 5.

21.    On Steward's views, see Chapter I above. Also see Boston Weekly Voice, October 18, 1866 (typescript, Wisconsin State Historical Society, Labor Collection, 11A, Box B); Edward Rogers in *Massachusetts Report on Hours*, 1867, p. 91; Massachusetts Bureau of Labor Statistics, *Fourth Annual Report*, 1873, pp. 449, 461, 470; Jesse H. Jones, "How to Attain the Eight Hour Day," *Gunton's Magazine*, XII (March 1897), 172; *The Garment Worker*, V (February 1901), 5–6.

22.    *Workingman's Advocate*, November 10, 1866, p. 2. Also see George Gunton, *Wealth and Progress* (New York, 1887), pp. 228–229. The socialists held the same position.

23.    *Workingman's Advocate*, June 30, 1866, p. 2; August 15, 1868, p. 2.

24.    Maine Bureau of Industrial and Labor Statistics, *Fourth Annual Report*, 1890, pp. 182–184. Also see Connecticut Bureau of Labor Statistics, *Second Annual Report*, 1886, p. xvii.

25.    New Jersey Bureau of Labor and Industrial Statistics, *Ninth Annual Report*, 1886, p. 233.

26.    Massachusetts Bureau of Labor Statistics, *Twelfth Annual Report*, 1881, pp. 458–462.

27.    William Gray, *Argument of Hon. William Gray on Petitions for Ten-Hour Law Before Committee on Labor, February 13, 1873* (Boston, 1873), pp. 17–18.

28.    Weekly American Workman, January 27, 1872, p. 1 (typescript in Wisconsin State Historical Society, Labor Collection, 11A).

29.  In the years before World War I, the leadership of the A. F. L. dropped the insistence that shorter hours not be linked to increased productivity. The decision reflected the failure to win eight hours on the old basis in many trades, and it was a pragmatic concession to the strength of the opposition. See *American Federationist* (June 1913), 461; (February 1914), 112-114; (May 1914), 325-326; (March 1916), 181-182; (September 1916), 841-842. Also see Jean T. McKelvey, *A. F. L. Attitudes Toward Production, 1900-1932* (Ithaca, N. Y., 1952).
30.  For an early statement of this position, see Horace Greeley, *New York Tribune*, July 12, 1865, p. 4.
31.  *Fincher's Trades' Review*, June 24, 1865, p. 29; September 2, 1865, p. 107; *Workingman's Advocate*, May 16, 1868, p. 2.
32.  *National Labor Tribune*, January 6, 1877, p. 2.
33.  *Workingman's Advocate*, July 1, 1871, p. 2.
34.  *Machinists' and Blacksmiths' International Journal*, XII (September 1875), 135.
35.  Ira Steward, "Meaning of the Eight Hour Movement," Ira Steward Papers, Wisconsin State Historical Society, Box 3.
36.  William Gray, *Argument Before Committee on Labor*, p. 11; Illinois Bureau of Labor Statistics, *Fourth Biennial Report*, 1886, p. 477; George Gunton, "Shall an Eight Hour System Be Adopted," *Forum*, I (April 1886), 143.
37.  Gunton, "Shall an Eight Hour System Be Adopted," p. 145.
38.  Maine Bureau of Industrial and Labor Statistics, *Third Annual Report*, 1889, pp. 45-46. The higher range in the locals that worked the fifty-three to fifty-four hour week was not explained.
39.  *Fincher's Trades' Review*, December 5, 1863, p. 3.
40.  *National Labor Tribune*, February 3, 1877, p. 1; *International Brotherhood of Stationary Firemen's Journal*, I (April 15, 1899), 7.
41.  *Fincher's Trades' Review*, September 9, 1865, p. 115.
42.  New York Bureau of Labor Statistics, *Third Annual Report*, 1885, p. 515.
43.  *The Worker*, I (February 9, 1873), 3; New York Bureau of Labor Statistics, *Eighth Annual Report*, 1890, pp. 535, 546; *The Tailor*, I (June 1889), 5; *The Carpenter* (March 1895), 8.
44.  *Labor Standard*, August 18, 1878, p. 4. Also see the *Labor Standard*, November 11, 1876, p. 3.
45.  *Painters' Journal*, V (November 1891), 5.
46.  *The Tailor*, I (May 1889), 1; III (January 1892), 2.
47.  Grob, *Workers and Utopia*, pp. 74-78.
48.  Knights of Labor, *Proceedings*, October 1886, pp. 40-41. Also see President L. R. Thomas in Pattern Makers' National League of North America, *Proceedings*, 1896, p. 8.
49.  American Federation of Labor, *Proceedings*, 1898, p. 122.
50.  Commons, *History of Labor*, II, p. 478.
51.  The carpenters were designated by the A. F. L. to achieve the eight-hour day by trade union action on May 1, 1890. In 1890, 36 cities were working eight hours. In 1892, carpenters worked eight hours in 46 cities, including New York, Baltimore, Brooklyn, Chicago, Denver, Indianapolis, Los Angeles, San Francisco, Seattle, Milwaukee, and St. Louis. By 1894, 54 cities had an eight-hour day. In 1896, the number increased to 70, in 1898 to 105 and in 1900 to 186. See United Brotherhood of Carpenters and Joiners of America, *Proceedings*, 1892, p. 15; 1894, p. 21; 1896, p. 28; 1898, p. 33; 1900, p. 48. On the response of the Carpenters' Union to machinery, see David Lyon, "The World of P. J. McGuire: A Study of the American Labor Movement, 1870-1890," Ph.D. dissertation, University of Minnesota, 1972, chs. 7, 9.
52.  United States, Labor Bureau, *Third Annual Report of the Commissioner of Labor* (Washington, D. C., 1887), pp. 16-17. In the period 1881-1886, strikes for reduced hours were successful in 24 percent of the cases, partially successful in 22 percent of the cases, and a failure 53 percent of the time. Comparable figures for

strikes over wage increases were 66 percent successful, 8 percent partly successful, and 25 percent failures. For all strikes during the 1881–1886 period, the figures were 46 percent successful, 13 percent partly successful, and 40 percent failures.

## CHAPTER X

1. *Workingman's Advocate*, November 2, 1867, p. 2. Also see Jonathan Grossman, *William Sylvis, Pioneer of American Labor: A Study of the Labor Movement During the Era of the Civil War* (New York, 1945), pp. 147, 281–282; Charlotte Erickson, *American Industry and the European Immigrant, 1860–1885* (Cambridge, 1957), pp. 51–52. For the efforts of employers in Pittsburgh to break strikes of puddlers by importing Belgian mechanics or English workers, see ibid., pp. 53-54.

2. *Workingman's Advocate*, September 28, 1867, p. 2.

3. Proceedings of the convention of the National Forge of the United Sons of Vulcan, August 1870, as found in the *Vulcan Record*, I, no. 6.

4. For the Workingmen's Assembly of the State of New York, see the *Workingman's Advocate*, February 5, 1870, p. 1, and February 12, 1870, p. 1; for the Coopers' International Union, see May 21, 1870, p. 2; for the Iron Molders' International Union, see July 23, 1870, p. 3, and August 13, 1870, p. 2; for the Carpenters' and Joiners' National Union, see October 8, 1870, p. 1; and for the Cigar Makers' International Union, see November 5, 1870, p. 1, and November 12, 1870, p. 1.

5. Ibid., January 22, 1870, p. 1.

6. *Welcome Workman*, February 22, 1868, p. 4. Also see *Workingman's Advocate*, June 13, 1868, p. 2. On the efforts to recruit workers and promote immigration from Europe, see Erickson, *American Industry and the European Immigrant*, chs. 1–4.

7. *Fincher's Trades' Review*, August 19, 1865, p. 96. On contacts between American and British trade unions, see Erickson, *American Industry and the European Immigrant*, pp. 54–60, and Grossman, *William Sylvis*, pp. 147–149.

8. *Fincher's Trades' Review*, September 2, 1865, p. 108.

9. Erickson, *American Industry and the European Immigrant*, pp. 61–62; *Workingman's Advocate*, November 2, 1867, p. 2.

10. National Labor Union, *Proceedings*, 1868, p. 38.

11. Minutes, March 11, 1872, Philadelphia Section, International Workingmen's Association, in International Workingmen's Association Papers, Wisconsin State Historical Society, Box 3. Mary Coolidge denied that the Chinese were imported labor in her *Chinese Immigration* (New York, 1909), pp. 41–54. On the "credit-ticket" system for Chinese immigrants, see Gunther Barth, *Bitter Strength: A History of the Chinese in the United States, 1850–1870* (Cambridge, 1964), pp. 55–56. Also see Ping Chiu, *Chinese Labor in California, 1850–1880: An Economic Study* (Madison, Wis., 1963).

12. On the characteristics of Chinese labor in California, see Alexander Saxton, *The Indispensable Enemy: Labor and the Anti-Chinese Movement in California* (Berkeley, 1971), pp. 1–10, 17. For examples of the arguments used against Chinese immigrants by specific trade unions, see for the Iron Molders' International Union, *Workingman's Advocate*, August 13, 1870, p. 2; for the Carpenters' and Joiners' National Union, ibid., October 8, 1870, p. 1; for the Cigar Makers' International Union, ibid., November 5, 1870, p. 1, and November 12, 1870, p. 1. Also see American Workman, August 13, 1870, p. 3 (typescript in Wisconsin State Historical Society, Labor Collection, 11A); *National Labor Tribune*, April 22, 1876, p. 2, and April 29, 1876, p. 2; *Labor Standard*, March 17, 1878, p. 2; April 14, 1878, p. 3; July 14, 1878, p. 1.

13. Barth, *Bitter Strength*, pp. 197–211. Also see *Workingman's Advocate*, June 12, 1869, p. 2.

14.   American Workman, August 13, 1870, p. 4 (typescript).
15.   For examples, see Workingman's Advocate, July 10, 1869, p. 2; July 17, 1869, p. 3; January 8, 1870, p. 2; September 24, 1870, p. 2; Coopers' Journal, IV (February 1873), 72; National Labor Tribune, May 27, 1876, p. 1. Coolidge argued that the hostility to Chinese workers in California was based upon racial antipathy. See: p. 378. Stuart C. Miller in The Unwelcome Immigrant: The American Image of the Chinese, 1785–1882 (Berkeley, 1969) stresses the racist nature of organized labor's attacks on the Chinese, and he ties this to the basically unfavorable image of the Chinese widely held in America.
16.   For a racist argument that demanded the restriction of immigration to the Caucasian race, further defined as "Celtic, Teutonic, Norman, and Saxon" peoples, see Hendrick Wright, A Practical Treatise on Labor (New York, 1871), ch. 8.
17.   Joseph Thompson, The Workman (New York, 1879), pp. 126–127. Barth argues that the American, Irish, and French-Canadian members of the Knights of St. Crispin did not initially respond with hostility to the Chinese workers introduced into North Adams, Massachusetts, in 1870 as strikebreakers. Instead they sought to organize a local lodge for Chinese Crispins. However, the effort failed, and Barth believes that this led Eastern workingmen into determined opposition to Chinese labor and thus to a policy of exclusion. See Bitter Strength, pp. 200–202. Saxton recognizes the importance of economic competition between the Chinese immigrants and other workers, but he stresses the importance of deep-seated ideological and psychological factors in explaining the hostility to the Chinese, which he likens more to the American reaction to its black population than to the nation's response to European immigrants. See Indispensable Enemy, particularly pp. 1–2, 17–44, 104–105, 121–122, 154–156, 258–284.
18.   United States Senate, Committee on Education and Labor, Report upon the Relations Between Labor and Capital, 4 vols. (Washington, D. C., 1885) I, pp. 791–792.
19.   American Federation of Labor, Proceedings, 1894, p. 12.
20.   Cigar Makers' Official Journal, XIX (April 1894), 8.
21.   United States Senate, Report upon Labor and Capital, I, pp. 520–521. This position modified George's statements in 1869, when he lived in California, which did include comments about the debased character and inferiority of the Chinese. See Saxton, Indispensable Enemy, pp. 102–103.
22.   Workingman's Advocate, October 2, 1869, p. 1. Also see February 12, 1870, p. 1.
23.   Massachusetts Bureau of Labor Statistics, Twelfth Annual Report, 1881, pp. 469–470.
24.   Massachusetts Bureau of Labor Statistics, Thirteenth Annual Report, 1882, pp. 88–89.
25.   Labor Standard, October 8, 1881, p. 4.
26.   See The Laster, April 15, 1891, p. 1; March 15, 1892, p. 2; May 15, 1892, p. 2. For a defense of the Armenian workers by the trade journal of the employers, see Boot and Shoe Recorder, December 9, 1891, p. 73. The Los Angeles Times argued in 1893 that "If we can keep out the Chinese, there is no good reason why we cannot exclude the lower classes of Poles, Hungarians, Italians and some other European nations, which people possess most of the vices of the Chinese and few of their good qualities, besides having a leaning towards bloodshed and anarchy which is peculiarly their own." Found in Saxton, Indispensable Enemy, p. 234.
27.   John Swinton in John Swinton's Paper, March 16, 1884, p. 1; William Weihe, United States Senate, Report upon Labor and Capital, II, p. 7.
28.   United States Senate, Report upon Labor and Capital, I, p. 338. On the Foran Act, see Erickson, American Industry and the European Immigrant, pp. 148–166.
29.   Federation of Organized Trades and Labor Unions of the United States and Canada, Proceedings, 1885, p. 8.
30.   For comments by workers, see Kansas Bureau of Labor and Industrial

Statistics, *First, Second, Third Annual Reports,* 1885, 1886, 1887; Iowa Bureau of Labor Statistics, *First Biennial Report,* 1884-1885; Colorado Bureau of Labor Statistics, *First Biennial Report,* 1887-1888; Michigan Bureau of Labor and Industrial Statistics, *Third Annual Report,* 1885; Ohio Bureau of Labor Statistics, *Thirteenth Annual Report,* 1889. In addition, several commissioners of labor statistics made clear their own support for a broader restriction of immigration. See Joel B. McCamant, Pennsylvania, Secretary of Internal Affairs, *Twelfth Annual Report,* Part III, 1884, p. 71; Charles Peck, New York Bureau of Labor Statistics, *Fourth Annual Report,* 1886 p. 61; E. R. Hutchins, Iowa Bureau of Labor Statistics, *First Biennial Report,* 1884-1885, pp. 184-185. Also see Wisconsin Bureau of Labor and Industrial Statistics, *Third Biennial Report,* 1887-1888.

31.     Kansas Bureau of Labor and Industrial Statistics, *First Biennial Report,* 1885, p. 104. Saxton sees a parallel between the racial factors so important in the hostility to the Chinese and the "complex strata of emotion and ideology" that overlay economic issues in the increasing opposition to southern and eastern Europeans. See *Indispensable Enemy,* pp. 273-278.

32.     Massachusetts Board to Investigate the Subject of the Unemployed, *Report,* Part V (Boston 1895), p. xxxviii.

33.     Knights of Labor, District Assembly no. 149, *Proceedings,* 1888, p. 30.

34.     Terence V. Powderly, "The Plea for Eight Hours," *North American Review,* CL (April 1890), 466; *Locomotive Firemen's Magazine,* XVI (March 1892), 253.

35.     William Godwin Moody, *Land and Labor in the United States* (New York, 1883), p. 275.

36.     John Simonds and John McEnnis, *The Story of Manual Labor in All Lands and Ages* (Chicago, 1886), p. 487.

37.     American Federation of Labor, *Proceedings,* 1896, pp. 81-82.

38.     On this point, see the comments by William Weihe, President of the Amalgamated Association of Iron and Steel Workers in United States Senate, *Investigation of Labor Troubles,* 52nd Congress, 2nd Session (Washington, D. C., 1893), pp. 215-216.

39.     Kansas Bureau of Labor and Industrial Statistics, *Sixth Annual Report,* 1890, p. 127; *The Railway Conductor,* XIII (September 1896), 559; *Iron Molders' Journal* (May 1896), pp. 194-195.

40.     American Federation of Labor, *Proceedings,* 1897, pp. 23, 56-57, 64, 88, 90-91.

41.     Ibid., 1889, p. 38.

42.     United Brotherhood of Carpenters and Joiners of America, *Proceedings,* 1896, p. 41.

43.     New York Bureau of Labor Statistics, *Sixteenth Annual Report,* 1898, p. 1030. Also see Glass Bottle Blowers' Association of the United States and Canada, *Proceedings,* 1896, p. 62.

44.     John White of the National Tobacco Workers' Union of America in American Federation of Labor, *Proceedings,* 1897, pp. 90-91.

45.     See E. Levasseur, *The American Workman* (Baltimore, 1900), p. 382, for his assessment of how strongly American workers feared competition from immigrants. Opponents of immigration outside the labor movement also recognized the workers' fear of competition from newcomers, and they incorporated this theme into their argument for restriction. For examples, see Richmond Mayo Smith, "Control of Immigration," *Political Science Quarterly,* III (June 1888), 223-224; Rena Michaels Atchinson, *Un-American Immigration: Its Present Effects and Future Perils* (Chicago, 1894), pp. 101-115; Francis A. Walker, "Immigration," *Yale Review,* I (August 1892), 136-137. Joseph M. Perry has argued that the substantial number of foreign-born workers in the cotton goods and iron and steel industries did produce lower wage levels than would have existed without immigration. See Joseph M. Perry, "The Impact of Immigration on Three American Industries, 1865-1914," Ph.D. dissertation, Northwestern University, 1966.

# BIBLIOGRAPHY

The bibliography is organized as follows:
I. Unpublished sources
II. Published sources
    1. Labor organizations
    2. Newspapers and periodicals
    3. Government publications
    4. Articles, books, and pamphlets by contemporaries
III. Books and articles

## I. UNPUBLISHED SOURCES

American Federation of Labor, Papers, Office of the President, Correspondence, 1888–1900, Wisconsin State Historical Society.

American Workman [1869–1872], typescript, Wisconsin State Historical Society, Labor Collection, 11A.

Arbitration between the Coal Operators and Miners in the Northern District of Illinois, Testimony, 1889, Wisconsin State Historical Society, Labor Collection, 14A, Box 2.

Bain, Trevor, The Impact of Technological Change on the Flat Glass Industry and the Unions' Reaction to Change: Colonial Period to the Present, Ph.D. dissertation, University of California, Berkeley, 1964.

Bemis, Edward, Papers, Wisconsin State Historical Society.

Cigar Makers' Society of the State of Maryland, Minutes, January 1856–June 1863, Johns Hopkins University.

Degler, Carl, Labor in the Economy and Politics of New York City, 1850–1860: A Study of the Impact of Early Industrialism, Ph.D. dissertation, Columbia University, 1952.

Faler, Paul, Workingmen, Mechanics and Social Change: Lynn, Massachusetts, 1800–1860, Ph.D. dissertation, University of Wisconsin, 1971.

Gutman, Herbert, Social and Economic Structure and Depression: American Labor in 1873 and 1874, Ph.D. dissertation, University of Wisconsin, 1959.

Hall, John P., The Gentle Craft: A Narrative of Yankee Shoemakers, Ph.D. dissertation, Columbia University, 1953.

Hat Finishers' Association of the City of Philadelphia and County of Camden, Journeymen Soft, Constitution and By-Laws, 1863, typescript, Wisconsin State Historical Society, Labor Collection 14A, Box 1.

Hat Finishers of the United States, National Trade Association of, Constitution, 1863, typescript, Wisconsin State Historical Society, Labor Collection 14A, Box 1.

International Workingmen's Association, Papers, Wisconsin State Historical Society.

Iron and Steel Heaters, Rollers, and Roughers of the United States, Associated Brotherhood of, Official Record of Correspondence, July 1874–July 1876, Wisconsin State Historical Society.

Iron Molders' International Journal, January–June 1874, typescript, Wisconsin State Historical Society, Labor Collection 14A, Box 1.

Knights of St. Crispin, Papers, Wisconsin State Historical Society.

Kugler, Israel, The Woman's Rights Movement and the National Labor Union, 1866–1872, Ph.D. dissertation, New York University, 1954.

Lasters' Union, Lynn, Papers, Baker Library, Harvard University.

Lyon, David N., The World of P. J. McGuire: A Study of the American Labor Movement, 1870–1890, Ph.D. dissertation, University of Minnesota, 1972.

McDonnell, Joseph P., Papers, Wisconsin State Historical Society.

Mule Spinners of New England, The Benevolent and Protective Association of the United Operative, Constitution and By-Laws, 1858, Wisconsin State Historical Society, Labor Collection 14A, Box 2.

Parsons, A. R., Papers, Wisconsin State Historical Society.

Perry, Joseph M., The Impact of Immigration on Three American Industries, 1865–1914, Ph.D. dissertation, Northwestern University, 1966.

Phillips, Thomas, Papers, Wisconsin State Historical Society.

Printers' Circular [February 1871–April 1872], typescript, Wisconsin State Historical Society, Labor Collection, 14A, Box 2.

Rogers, Edward, Papers, Wisconsin State Historical Society.

Samuel, John, Papers, Wisconsin State Historical Society.

Schilling, Robert, Papers, Wisconsin State Historical Society.

Sovereigns of Industry, Papers, Wisconsin State Historical Society.

Steward, Ira, Papers, Wisconsin State Historical Society.

Typographical Union, National, Synopsis of Proceedings, 1853–1856, typescript of The Printer, January 1860–March 1860, Wisconsin State Historical Society.

Union Cooperative Association, Papers, Wisconsin State Historical Society.

Union Cooperative Printing Company (St. Louis), Minutes, 1880–1881, Wisconsin State Historical Society.

## II. PUBLISHED SOURCES

### 1. Labor Organizations

Amalgamated Association of Iron and Steel Workers of the United States, Proceedings, 1876-1895.

Amalgamated Building Trades' Council (New York), Constitution and By-Laws, 1885.

American Federation of Labor, Proceedings, 1886-1901.

American Federationist, 1894-1900, 1913-1916.

American Flint Glass Workers' Union, History [of] American Flint Glass Workers' Union of North America [Toledo], 1957.

————, Proceedings, 1887-1903.

Boot and Shoe Workers' Union, Proceedings, 1895-1899.

*Bricklayer and Mason, The,* 1898–1900.
Bricklayers' and Masons' International Union, *Proceedings,* 1885–1900.
Bricklayers' National Union, *Proceedings,* 1865–1868.
*Brotherhood of Locomotive Engineers Monthly Journal,* 1873, 1882–1889, 1894.
*Carpenter, The,* 1891–1900.
Carpenters' and Joiners' National Union of the United States of America, *Proceedings,* 1865–1867.
*Carriage and Wagon Workers' Journal,* 1899.
*Cigar Makers' Official Journal,* 1876–1900.
Connecticut Federation of Labor, *Proceedings,* 1887, 1892.
*Coopers' International Journal,* 1900–1904.
Coopers' International Union, *Constitution,* 1897–1908.
*Coopers' Journal,* 1870–1873.
Federation of Organized Trades and Labor Unions of the United States and Canada, *Proceedings,* 1881–1886.
*Foremen's Advance Advocate, The,* 1892–1896.
Furniture Workers' Union of North America, Central Committee, *Normal Workday of Eight Hours* (New York, 1879).
*Garment Worker, The,* 1895–1901.
Glass Bottle Blowers' Association of the United States and Canada, *Proceedings,* 1892–1906.
Granite Cutters' National Union of the United States, *Constitution and By-Laws,* 1880, 1888, 1896.
*Hammer, The* (Metal Workers' Union of North America), 1882–1887.
Illinois Miners' Protective Association, *Proceedings,* 1889.
*International Brotherhood of Stationary Firemen's Journal,* 1899.
*International Journal, The* [of the Iron Molders' International Union], 1866–1867.
*International Stove Mounters' Journal,* 1898–1901.
International Typographical Union, *Proceedings,* 1857–1893 (1894–1900 in *Typographical Journal*).
Iron and Steel Roll Hands' Union of the United States, *Proceedings,* 1873–1874, 1876.
*Iron Molders' Journal* [1875], 1895–1900.
Joint Conference of Miners and Operators, Interstate Convention, *Proceedings,* 1898–1912.
*Journal, The* (Metal Polishers, Buffers, Platers and Brass Workers' Union of North America), 1897–1899.
*Journal of United Hatters of North America,* 1898–1900.
*Journal of United Labor* (Knights of Labor), 1880–1889.
Journeymen Cigar Makers' Union of the United States, *Proceedings,* 1864–1867 [1870].
Journeymen Stone Cutters' Association of North America, *Constitution and By-Laws,* 1887, 1892, 1900–1929.
Journeymen Stone-Cutters' Association of the District of Columbia, *Constitution and By-Laws,* 1854.
Journeymen Tailors' Union of America, *Constitution,* 1887–1921.
Knights of Labor, *Proceedings,* 1878–1896.
Knights of Labor, District Assembly No. 149, *Proceedings,* 1887–1889.
Knights of St. Crispin, Hanson Lodge No. 135, *Constitution and By-Laws,* 1869.
*Laster, The* (Lasters' Protective Union of America), 1888–1892.
Lasters' Protective Union of America (New England Lasters' Protective Union to 1890), *Proceedings,* October 1888, April 1889, 1895.
*Leather Workers' Journal, The,* 1898–1900.
*Locomotive Firemen's Magazine,* 1886–1899.
*Machine Wood-Worker, The,* 1890–1899 (*American Wood-Worker,* January–April 1895; *International Wood-Worker,* 1896–1899).

*Machinery Molders' Journal,* 1888–1892.
*Machinists' and Blacksmiths' International Journal, The,* 1870–1872 [1875].
Machinists' and Blacksmiths' Union, *Ritual, Containing the Initiatory Ceremonies; Rules Concerning Strikes; Rate of Relief to Members, Etc.,* 1876.
*Machinists' Monthly Journal,* 1899–1903.
Mechanics State Council [California], *Constitution and By-Laws with Names of Associations and Delegates: A Condensed History of the Eight-Hour Movement in California; Corrected Copies of Reform Acts of the Legislature Passed in 1868* [and] *the Constitution of the National Labor Union,* 1868.
National Federation of Miners and Mine Laborers of the United States and Territories, *Proceedings,* 1886–1888.
National Labor Union, *Address to the People of the United States on Money, Land and Other Subjects of National Importance,* 1870.
————, *Proceedings,* 1868.
National Progressive Union of Miners and Mine Laborers, *Proceedings,* 1888.
New England Lasters' Protective Union, *Prices Paid for Lasting Boots and Shoes in Maine, New Hampshire and Massachusetts,* April 1887.
New York State Workingmen's Assembly, *Proceedings,* 1869–1871.
Order of Railway Conductors, *Proceedings,* 1868–1885.
*Painters' Journal (The Painter* to December 1889), 1887–1895.
*Pattern Makers' Monthly Journal,* 1893–1898.
Pattern Makers' National League of North America, *Proceedings,* 1888–1900.
*Plumbers, Gas and Steam Fitters' Journal,* 1900–1901.
*Railway Conductor, The,* 1890–1900.
*Stone Cutters' Journal,* 1896–1903.
*Tailor, The,* 1888–1900.
*Trackmen's Advance Advocate,* 1897–1900.
*Typographical Journal, The,* 1889–1900.
United Association of Journeymen Plumbers, Gas Fitters, Steam Fitters and Steam Fitters' Helpers, *Proceedings,* 1897.
United Brotherhood of Carpenters and Joiners of America, *Constitution,* 1881, 1886, 1888–1898.
————, *Proceedings,* 1888–1900.
United Mine Workers of America, *Proceedings,* 1899–1908.
*Vulcan Record,* 1868–1875.
Window Glass Workers, Local Assembly 300, Knights of Labor, *Proceedings,* 1884, 1889, 1892, 1895, 1896, 1899.

## 2. Newspapers and Periodicals

*Boot and Shoe Recorder,* 1890–1891.
*Equity,* 1874–1875.
*Fincher's Trades' Review,* 1863–1865.
*Iron Age, The,* 1876–1879.
*John Swinton's Paper,* 1883–1887.
*Justice,* 1883–1885.
*Labor-Balance, The,* 1877–1878.
*Labor Standard,* 1876–1881.
*Milwaukee Sentinel,* May 1–15, 1886.
*National Labor Tribune,* 1875–1878.
*National Workman, The,* 1866–1867.
*New York Times,* June–August 1868.
*New York Tribune,* June–August 1868, May–June 1872.
*Socialist, The,* 1876.
*Toiler, The,* 1874.

*Welcome Workman,* 1867–1868.
*Word, The,* 1872–1879.
*Work and Wages,* 1886–1887.
*Worker, The,* 1873.
*Workingman's Advocate,* 1864–1877.

## 3. Government Publications

California, Bureau of Labor Statistics, *Biennial Reports,* 1–5, 1883–1892.
Colorado, Bureau of Labor Statistics, *Biennial Reports,* 1–5, 7, 1887–1896, 1899–1900.
Connecticut, Bureau of Labor Statistics, *Annual Reports,* 1–2, 1874–1875; second series, 1–15, 1885–1899.
Illinois, Bureau of Labor Statistics, *Biennial Reports,* 1–11, 1881–1900.
Illinois, Bureau of Labor Statistics, *Annual Reports on Coal,* 12–19, 1893–1900.
Indiana, Statistics Bureau, *Annual Reports,* 4–5, 1882–1883; *Biennial Reports,* 7, 10–11, 13, 1885–1886, 1891–1894, 1897–1898.
Iowa, Bureau of Labor Statistics, *Biennial Reports,* 1–7, 1884–1896.
Kansas, Bureau of Labor and Industrial Statistics, *Annual Reports,* 1–6, 11–12, 14, 1885–1890, 1895–1896, 1898.
Maine, Bureau of Industrial and Labor Statistics, *Annual Reports,* 1–4, 1887–1890.
Maryland, Bureau of Industrial Statistics, *Biennial Reports,* 1, 3–4, 1884–1885, 1888–1891; *Annual Reports,* 1–3, 5–6, 8–9, 1892–1894, 1896–1897, 1899–1900.
Massachusetts Board to Investigate the Subject of the Unemployed, *Report,* 1895.
Massachusetts, Bureau of Labor Statistics, *Annual Reports,* 1–30, 1870–1900.
Massachusetts, Bureau of Labor Statistics, *Census of Massachusetts,* 1875, 1885.
Michigan, Bureau of Labor and Industrial Statistics, *Annual Reports,* 1–17, 1883–1899.
Minnesota, Bureau of Labor Statistics, *Biennial Reports,* 1–4, 7, 1887–1894, 1899–1900.
Missouri, Bureau of Labor Statistics, *Annual Reports,* 1–4, 14–15, 17, 20–22, 1879–1882, 1892–1893, 1895, 1898–1900.
Nebraska, Bureau of Labor and Industrial Statistics, *Biennial Report,* 1, 1887–1888.
New Jersey, Bureau of Labor and Industrial Statistics, *Annual Reports,* 1–23, 1878–1900.
New York, Bureau of Labor Statistics, *Annual Reports,* 1–17, 1883–1899.
Ohio, Bureau of Labor Statistics, *Annual Reports,* 1–24, 1877–1900.
Pennsylvania, Secretary of Internal Affairs, *Annual Reports: Industrial Statistics,* 1–28, 1872/73–1900.
*Report of the Special Commission on the Hours of Labor and the Condition and Prospects of the Industrial Classes* (Massachusetts House, No. 98), 1866.
*Reports of Commissioners on the Hours of Labor* (Massachusetts House, No. 44), 1867.
United States, Coal Commission, *Report,* Part V, 1925.
United States, House of Representatives, *Investigation by a Select Committee of the House of Representatives Relative to the Causes of the General Depression in Labor and Business; And as to Chinese Immigration* (46th Congress, 2nd Session, Miscellaneous Document No. 5), 1879.
United States, House of Representatives, *Labor Troubles at Homestead, Pennsylvania* (52nd Congress, 2nd Session, Report No. 244), 1893.
United States, House of Representatives, *Labor Troubles in the Anthracite Regions of Pennsylvania, 1887–1888* (50th Congress, 2nd Session, Report No. 4147), 1889.
United States Industrial Commission, *Report,* 19 vols., 1900–1902.
United States, Labor Bureau, *Bulletin* No. 13 (G. O. Virtue, "The Anthracite Mine Laborers," pp. 728–774), 1897.
United States, Labor, Commissioner of, *Annual Reports,* 1–13, 1886–1898.
————, Eleventh Special Report, *Regulation and Restriction of Output,* 1904.

United States, Labor, Department of, Bureau of Labor Statistics, *Bulletin* No. 441 (Boris Stern, "Productivity of Labor in the Glass Industry"), July 1927.

United States National Resources Committee, *Technological Trends and National Policy* (Bernhard J. Stern, "Resistances to the Adoption of Technological Innovations," pp. 39–66), 1937.

United States, Senate, *Investigation of Labor Troubles* (52nd Congress, 2nd Session, Report No. 1280), 1893.

United States, Senate, Committee on Education and Labor, *Report upon the Relations Between Labor and Capital*, 4 vols., 1885.

United States, Works Progress Administration, National Research Project on Employment Opportunities, *Report L-1* (Daniel Creamer and Gladys Swackhamer, "Cigar Makers–After the Lay-Off: A Case Study of the Effects of Mechanization on Employment of Hand Cigar Makers"), 1937.

————, *Report L-8* (Harry Ober, "Trade Union Policy and Technological Change"), 1940.

United States, Works Progress Administration, National Research Project on Reemployment Opportunities and Recent Changes in Industrial Techniques, *Report B-4* (W. D. Evans, "Effects of Mechanization in Cigar Manufacture"), 1938.

Wisconsin, Bureau of Labor and Industrial Statistics, *Biennial Reports*, 2–3, 1885–1888.

## 4. Articles, Books, and Pamphlets by Contemporaries

"Among the Nail-Makers," *Harper's New Monthly Magazine*, XXI (July 1860), 145–164.

Atchinson, Rena Michaels, *Un-American Immigration: Its Present Effects and Future Perils* (Chicago, 1894).

Atwater, Lyman, "The Labor Question in Its Economic and Christian Aspects," *Presbyterian Quarterly and Princeton Review*, n. s. I (1872), 468–495.

Bemis, Edward, "The Homestead Strike," *Journal of Political Economy*, II (June 1894), 369–396.

————, "Relation of Labor Organizations to the American Boy and to Trade Instruction," *The Annals*, American Academy of Political and Social Science, V (September 1894), 209–241.

Bigelow, Erastus, "The Relations of Labor and Capital," *Atlantic Monthly*, XLI (October 1878), 475–487.

Cameron, A. S., *The Eight Hour Question* (New York, 1872).

Clark, J. B., "The Labor Problem–Past and Present," *Work and Wages*, I (January 1887).

Commons, John R., "The Right to Work," *The Arena*, XXI (February 1899), 131–142.

"Comparison of Hand and Machine Production," *Engineering News*, XLII (October 5, 1899), 224–225.

Davis, C. Wood, "Does Machinery Displace Labor?" *Forum*, XXV (July 1898), 603–617.

"Destiny of the Mechanic Arts," *Harper's New Monthly Magazine*, XIV (April 1857), 696–699.

Dewey, Davis R., "Irregularity of Employment," American Economic Association, *Publications*, IX (1894), 53–67.

"The Eight-Hour Delusion," *The Nation*, III (November 22, 1866), 412–413.

"The Eight-Hour Working-day," *The Century*, XXXIII (December 1886), 318–320.

Eliot, Samuel, "Relief of Labor," *Journal of Social Science*, IV (1871), 133–149.

Ely, Richard, *The Labor Movement in America* (New York, 1886).

Flower, B. O., "Emergency Measures Which Would Have Maintained Self-Respecting Manhood," *The Arena*, IX (April 1894), 822–826.

George, Henry, "Overproduction," *North American Review,* CXXXVII (December 1883), 584–593.

———, *Social Problems* (Chicago, 1883).

Gladden, Washington, "What to Do with the Workless Man," National Conference of Charities and Correction, *Proceedings* (1899), 141–152.

———, *Working People and Their Employers* (New York, 1885).

Godkin, Edward, "The Causes of the Industrial Depression," *The Nation,* XXVII (October 3, 1878), 206–207.

Going, Charles, "Labor Questions in England and America," *The Engineering Magazine,* XIX (May 1900), 165–172.

Graffenreid, Claire de, "Child Labor," American Economic Association, *Publications,* V (1890), 193–271.

Gray, William, *Argument of Hon. William Gray on Petitions for Ten-Hour Law Before Committee on Labor, February 13, 1873* (Boston, 1873).

Gregory, John M., "The Problem of the Unemployed," *The Independent,* XXXIX (November 10, 1887), 1443.

[Gunton, George], "Does Invention Lessen Employment?" *Gunton's Magazine,* XIV (May 1898), 331–338.

Gunton, George, *The Economic and Social Importance of the Eight-Hour Movement* (American Federation of Labor, Eight Hour Series, No. 2, Washington, D. C., 1889).

———, "Shall an Eight Hour System Be Adopted," *Forum,* I (April 1886), 136–148.

———, *Wealth and Progress* (New York, 1887).

Hadley, Arthur T., *Economics: An Account of the Relations Between Private Property and Public Welfare* (New York, 1896).

Halsey, F. A., "Experience with the Premium Plan of Paying for Labor," *American Machinist,* XXII (March 9, 1899), 17–20.

Harris, William, "Is There Work Enough for All?" *Forum,* XXV (April 1898), 224–236.

Hewitt, Edward, *Remarkable Developments in the United States: Un-Paralleled Activities, 1886–1892* (New York, 1898).

Hobson, John, "The Influence of Machinery upon Employment," *Political Science Quarterly,* VIII (March 1893), 97–123.

Jelley, S. M., *The Voice of Labor* (Philadelphia, 1888).

Jenks, Jeremiah, "A Word to Trades-Unions," *Charities Review,* I (December 1891), 55–59.

*Johns Hopkins University Studies in Historical and Political Science,* VI (1888).

Jones, Jesse H., "How to Attain the Eight Hour Day," *Gunton's Magazine,* XII (March 1897), 169–175.

Kellogg, D. O., *Thoughts on the Labor Question* (1879).

"The Labor Crisis," *The Nation,* IV (April 25, 1867), 334–336.

*Labor: Its Rights and Wrongs* (Washington, D. C., 1886).

"Labor Strikes," *Harper's New Monthly Magazine,* XLVII (June 1873), 142–143.

Levasseur, E., *The American Workman* (Baltimore, 1900).

Logan, Walter, *An Argument for an Eight-Hour Law* (New York, 1894).

McNeill, George, *The Eight Hour Primer* (American Federation of Labor, Eight Hour Series, No. 1, Washington, D. C., 1888).

——— (ed.), *The Labor Movement: The Problem of To-Day* (New York, 1888).

———, "The Struggle for Life," *Work and Wages,* I (November 1886).

Mavor, James, "Labor Colonies and the Unemployed," *Journal of Political Economy,* II (1893–1894), 26–53.

Maxim, Hiram, "Automatic Machinery: The Secret of Cheap Production," *The Engineering Magazine,* XIV (January 1898), 593–602.

———, "The Effects of Trade Unionism Upon Skilled Mechanics," *The Engineering Magazine,* XIV (November 1897), 189–195.

Means, D. Mcgregor, "The Dangerous Absurdity of State Aid," *Forum,* XVII (May 1894) 287–296.
Moody, William Godwin, "Foreign Trade No Cure for Hard Times," *Atlantic Monthly,* XLIV (October 1879), 472–476.
————, *Land and Labor in the United States* (New York, 1883).
————, *Our Labor Difficulties: The Cause and the Way Out* (Boston, 1878).
————, *A Paper on the Displacement of Labor by Improvements in Machinery* (Boston, 1878).
"The Multiplication of Industries," *Scribner's Monthly,* XIII (April 1877), 863–864.
National Guard of Industry, *Platform and Subordinate Constitution* (Washington, D. C., 1870).
Newcomb, Simon, *A Plain Man's Talk on the Labor Question* (New York, 1886).
New England Labor-Reform League, *Declaration of Sentiments and Constitution* (Boston, 1869).
"New Shop Methods: A Corollary of Modern Machinery," *The Engineering Magazine,* XIX (June 1900), 369–372.
Newton, R. Heber, *The Present Aspect of the Labor Problem* (New York, 1886).
Nichols, Starr H., "Men and Machinery," *North American Review,* CLXVI (May 1898), 602–611.
Norris, H. M., "Actual Experience with the Premium Plans," *The Engineering Magazine,* XVIII (January 1900), 572–584; (February 1900), 689–696.
O'Connell, James, "Length of Trade Life Amongst Machinists," *The Annals,* American Academy of Political and Social Science, XXVII (May 1906), 491–495.
————, "Piece-Work Not Necessary for Best Results in the Machine Shop," *The Engineering Magazine,* XIX (June 1900), 373–380.
Outerbridge, A. E. Jr., "Labor-Saving Machinery: The Secret of Cheap Production," *The Engineering Magazine,* XII (January 1897), 650–656.
Peters, Edward, "Some Economic and Social Effects of Machinery," American Association for the Advancement of Science, *Proceedings* (1884), 638–642.
Powderly, Terence V., "The Plea for Eight Hours," *North American Review,* CL (April 1890), 464–469.
————, *Thirty Years of Labor* (Columbus, Ohio, 1889).
Richardson, A. D., "Making Watches by Machinery," *Harper's New Monthly Magazine,* XXXIX (July 1869), 169–182.
Sanial, Lucien, *Socialist Almanac and Treasury of Facts* (New York, 1898).
Shaw, Albert, "Cooperation in a Western City," American Economic Association, *Publications,* I (September 1886).
Simonds, John and McEnnis, John, *The Story of Manual Labor in All Lands and Ages* (Chicago, 1886).
Smith, Richmond Mayo, "Control of Immigration," *Political Science Quarterly,* III (1888), 46–77, 197–225, 409–424.
————, "The National Bureau of Labor and Industrial Depressions," *Political Science Quarterly,* I (September 1886), 437–448.
Spahr, Charles, "The Taxation of Labor: The American Theory," *Political Science Quarterly,* I (September 1886), 400–436.
Strong, Josiah, *The New Era; or the Coming Kingdom* (New York, 1893).
Sylvis, James (ed.), *The Life, Speeches, Labors and Essays of William H. Sylvis* (Philadelphia, 1872).
Thompson, Joseph, *The Workman* (New York, 1879).
"The True Cause of the Hard Times," *The Nation,* XL (February 26, 1885), 179.
"The True History of the Coal Trouble," *The Nation,* XII (March 9, 1871), 152–154.
Tuttle, Charles, "The Workman's Position in the Light of Economic Progress," American Economic Association, *Papers and Proceedings of the Fourteenth Annual Meeting,* December 1901, 199–234.
Versteeg, Dingman, *Labor Saving Machinery and Progress* (New York, 1895).
Walker, Francis A., "Immigration," *Yale Review,* I (August 1892), 125–145.

Wells, David, "The Influence of the Production and Distribution of Wealth on Social Development," *Journal of Social Science*, VIII (1876), 1–22.

White, Henry, "Machinery and Labor," *The Annals*, American Academy of Political and Social Science, XX (July 1902), 223–231.

Willoughby, William F., "Child Labor," American Economic Association, *Publications*, V (March 1890), 129–192.

————, *Workingmen's Insurance* (New York, 1898).

Wright, Carroll, "The Amalgamated Association of Iron and Steel Workers," *Quarterly Journal of Economics*, VII (July 1893), 400–432.

————, "Are the Rich Growing Richer and the Poor Poorer?" *Atlantic Monthly*, LXXX (September 1897), 300–309.

————, "The Relation of Production to Productive Capacity," *Forum*, XXIV (November 1897), 290–302; (February 1898), 660–675.

Wright, Hendrick, *A Practical Treatise on Labor* (New York, 1871).

Young, Edward, *Labor in Europe and America* (Philadelphia, 1875).

## III. BOOKS AND ARTICLES

Aurand, Harold, *From the Molly Maguires to the United Mine Workers: The Social Ecology of an Industrial Union, 1869–1897* (Philadelphia, 1971).

Baer, Willis, *The Economic Development of the Cigar Industry in the United States* (Lancaster, Pa., 1933).

Baker, Elizabeth Faulkner, *Displacement of Men by Machines: Effects of Technological Change in Commercial Printing* (New York, 1933).

————, *Printers and Technology: A History of the International Printing Pressmen and Assistant's Union* (New York, 1957).

Baratz, Morton, *The Union and the Coal Industry* (New Haven, 1955).

Barlow, Melvin, *History of Industrial Education in the United States* (Peoria, 1967).

Barnett, George, *Chapters on Machinery and Labor* (Cambridge, 1926).

Barth, Gunther, *Bitter Strength: A History of the Chinese in the United States, 1850–1870* (Cambridge, 1964).

Bates, Harry, *Bricklayers' Century of Craftsmanship: A History of the Bricklayers, Masons and Plasterers' International Union of America* (Washington, D. C., 1955).

Bedford, Henry, *Socialism and the Workers in Massachusetts, 1886–1912* (Amherst, 1966).

Blumin, Stuart, "Mobility and Change in Ante-Bellum Philadelphia," in Stephan Thernstrom and Richard Sennett (eds.), *Nineteenth-Century Cities: Essays in the New Urban History* (New Haven, 1969).

Bonnett, Clarence, *History of Employers' Associations in the United States* (New York, 1956).

Bremner, Robert, *From the Depths: The Discovery of Poverty in the United States* (New York, 1956).

Briggs, Asa, *The Age of Improvement, 1783–1867* (New York, 1959).

———— (ed.), *Chartist Studies* (New York, 1959).

Brody, David, *The Butcher Workmen: A Study of Unionization* (Cambridge, 1964).

————, *Steelworkers in America: The Non-Union Era* (Cambridge, 1960).

Broehl, Wayne, *The Molly Maguires* (Cambridge, 1964).

Cahill, Marion, *Shorter Hours: A Study of the Movement Since the Civil War* (New York, 1932).

Cawelti, John, *Apostles of the Self-Made Man* (Chicago, 1965).

Chiu, Ping, *Chinese Labor in California, 1850–1880: An Economic Study* (Madison, Wis., 1963).

Christie, Robert, *Empire in Wood: A History of the Carpenters' Union* (Ithaca, N. Y., 1956).

Cole, Arthur, and Williamson, Harold, *The American Carpet Manufacture* (Cambridge, 1941).

Commager, Henry, *The American Mind: An Interpretation of American Thought and Character Since the 1880's* (New Haven, 1950).

Commons, John, et al., *A Documentary History of American Industrial Society,* 10 vols. (Cleveland, 1910).

————, *History of Labor in the United States,* Vols. 1–2 (New York, 1918); Vols. 3–4 (New York, 1935).

Coolidge, Mary, *Chinese Immigration* (New York, 1909).

Coyne, Franklin, *The Development of the Cooperage Industry in the United States, 1620–1940* (Chicago, 1940).

Cremin, Lawrence, *The Transformation of the School: Progressivism in American Education, 1876–1957* (New York, 1961).

Curoe, Philip, *Educational Attitudes and Policies of Organized Labor in the United States* (New York, 1926).

Davis, Allen F., *Spearheads of Reform: The Social Settlements and the Progressive Movement, 1890–1914* (New York, 1967).

Davis, Pearce, *The Development of the American Glass Industry* (Cambridge, 1949).

Deibler, Frederick, *The Amalgamated Wood Workers' International Union of America* (Madison, 1912).

Dorfman, Joseph, *The Economic Mind in American Civilization,* Vol. III: 1865–1918 (New York, 1949).

Douglas, Dorothy, "Ira Steward on Consumption and Unemployment," *Journal of Political Economy,* XL (1932), 532–543.

Erickson, Charlotte, *American Industry and the European Immigrant, 1860–1885* (Cambridge, 1957).

Evans, Chris, *History of United Mine Workers of America,* Vol. 1: 1860–1890; Vol. 2: 1890–1900 [1918].

Feder, Leah, *Unemployment Relief in Periods of Depression: A Study of Measures Adopted in Certain American Cities, 1857 through 1922* (New York, 1936).

Felt, Jeremy, *Hostages of Fortune: Child Labor Reform in New York State* (Syracuse, 1965).

Fine, Sidney, "The Eight-Hour Day Movement in the United States 1888–1891," *Mississippi Valley Historical Review,* XL (December 1953), 441–462.

————, *Laissez-Faire and the General-Welfare State: A Study of Conflict in American Thought, 1865–1901* (Ann Arbor, Mich., 1956).

Fisher, Berenice, *Industrial Education: American Ideals and Institutions* (Madison, 1967).

Fisher, Waldo and Bezanson, Anne, *Wage Rates and Working Time in the Bituminous Coal Industry, 1912–22* (Philadelphia, 1932).

Fite, Emerson, *Social and Industrial Conditions in the North During the Civil War* (New York, 1910).

Flexner, Eleanor, *Century of Struggle: The Women's Rights Movement in the United States* (Cambridge, 1959).

Foner, Philip, *History of the Labor Movement in the United States,* Vols. 1, 2 (New York, 1947, 1955).

Galombos, Louis, "A. F. L.'s Concept of Big Business: A Quantitative Study of Attitudes Toward the Large Corporation, 1894–1931," *Journal of American History,* LVII (March 1971), 847–863.

Green, Charles, *The Headwear Workers: A Century of Trade Unionism* (New York, 1944).

Greenbaum, Fred, "The Social Ideas of Samuel Gompers," *Labor History,* VII (Winter 1966), 35–61.

Greene, Victor, *The Slavic Community on Strike: Immigrant Labor in Pennsylvania Anthracite* (Notre Dame, Ind., 1968).

Grob, Gerald, *Workers and Utopia: A Study of Ideological Conflict in the American Labor Movement, 1865-1900* (Evanston, Ill., 1961).

Grossman, Jonathan, *William Sylvis, Pioneer of American Labor: A Study of the Labor Movement During the Era of the Civil War* (New York, 1945).

Gutman, Herbert, "Labor's Response to Modern Industrialism," in Howard Quint et al., *Main Problems in American History*, Vol. II (Homewood, Ill., 1964), 71-95.

———, "The Reality of the Rags-to-Riches 'Myth': The Case of the Paterson, New Jersey, Locomotive, Iron, and Machinery Manufacturers, 1830-1880," in Stephan Thernstrom and Richard Sennett (eds.), *Nineteenth-Century Cities: Essays in the New Urban History* (New Haven, 1969).

———, "The Worker's Search for Power: Labor in the Gilded Age," in H. Wayne Morgan (ed.), *The Gilded Age: A Reappraisal* (Syracuse, 1963).

Haber, William, *Industrial Relations in the Building Industry* (Cambridge, 1930).

Hall, John P., "The Knights of St. Crispin in Massachusetts, 1869-1878," *Journal of Economic History*, XVIII (June 1958), 161-175.

Higham, John, *Strangers in the Land: Patterns of American Nativism, 1860-1925* (New Brunswick, 1955).

Hoagland, Henry, *Collective Bargaining in the Lithographic Industry* (New York, 1917).

Hobsbawm, E. J., "Custom, Wages and Work-Load in Nineteenth Century Industry," in *Labouring Men* (Garden City, 1967), 405-435.

———, "The Machine Breakers," in *Labouring Men* (Garden City, 1967), 7-26.

Hugins, Walter, *Jacksonian Democracy and the Working Class: A Study of the New York Workingmen's Movement, 1829-1837* (Stanford, 1960).

Jerome, Harry, *Mechanization in Industry* (New York, 1934).

Karson, Marc, *American Labor Unions and Politics, 1900-1918* (Carbondale, 1958).

Korman, Gerd, *Industrialization, Immigrants and Americanizers: The View from Milwaukee, 1866-1921* (Madison, 1967).

Laslett, John, *Labor and the Left: A Study of Socialist and Radical Influences in the American Labor Movement, 1881-1924* (New York, 1970).

Lebergott, Stanley, "Wage Trends, 1800-1900," in National Bureau of Economic Research, *Trends in the American Economy in the Nineteenth Century* (Princeton, 1960), 449-499.

Leiby, James, *Carroll Wright and Labor Reform* (Cambridge, 1960).

Lescohier, Don, *The Knights of St. Crispin: A Study in the Industrial Causes of Trade Unionism* (Madison, 1910).

Long, Clarence, *Wages and Earnings in the United States, 1860-1890* (Princeton, 1960).

Lubin, Isador, *Miners' Wages and the Cost of Coal* (New York, 1924).

Lubove, Roy, *The Struggle for Social Security, 1900-1935* (Cambridge, 1968).

McCabe, David, *The Standard Rate in American Trade Unions* (Baltimore, 1912).

Mack, Russell, *The Cigar Manufacturing Industry: Factors of Instability Affecting Production and Employment* (Philadelphia, 1933).

McKelvey, Jean, *A. F. L. Attitudes Toward Production, 1900-1932* (Ithaca, 1952).

Mandel, Bernard, *Samuel Gompers* (Yellow Springs, 1963).

Martin, Edgar, *The Standard of Living in 1860: American Consumption Levels on the Eve of the Civil War* (Chicago, 1942).

Marx, Leo, *The Machine in the Garden* (New York, 1964).

Mathewson, Stanley, *Restriction of Output among Unorganized Workers* (New York, 1931).

Mayer, Thomas, "Some Characteristics of Union Members in the 1880's and 1890's," *Labor History*, V (Winter 1964), 57-66.

Miller, Stuart C., *The Unwelcome Immigrant: The American Image of the Chinese, 1785-1882* (Berkeley, 1969).

Minton, Lee, *Flame and Heart: A History of the Glass Bottle Blowers' Association of the United States and Canada* (1961).

Mohl, Raymond, *Poverty in New York, 1783-1825* (New York, 1971).

Mohr, James, *The Radical Republicans and Reform in New York During Reconstruction* (Ithaca, 1973).

Montgomery, David, *Beyond Equality: Labor and the Radical Republicans, 1862–1872* (New York, 1967).

————, "The Working Classes of the Pre-Industrial American City, 1780–1830," *Labor History*, IX (Winter 1968), 3–22.

Morison, Elting, *Men, Machines and Modern Times* (Cambridge, 1966).

Motley, James, *Apprenticeship in American Trade Unions* (Baltimore, 1907).

Mowry, George, *The Era of Theodore Roosevelt, 1900–1912* (New York, 1958).

Nadworny, Milton, *Scientific Management and the Unions, 1900–1932* (Cambridge, 1955).

Nelson, Daniel, *Unemployment Insurance: The American Experience, 1915–1935* (Madison, 1969).

Ozanne, Robert, *A Century of Labor-Management Relations at McCormick and International Harvester* (Madison, 1967).

Palmer, Gladys, *Union Tactics and Economic Change* (Philadelphia, 1932).

Panschar, William, *Baking in America: Economic Development* (Evanston, 1956).

Patterson, Joseph, "The Old W. B. A. Days," Historical Society of Schuylkill County, *Publications*, II (1910), 355–384.

Perlman, Mark, *The Machinists: A New Study in American Trade Unionism* (Cambridge, 1961).

Perlman, Selig, *A Theory of the Labor Movement* (New York, 1928).

Pessen, Edward, "The Egalitarian Myth and the American Social Reality: Wealth, Mobility and Equality in the 'Era of the Common Man,'" *American Historical Review*, LXXVI (October 1971), 989–1034.

————, *Most Uncommon Jacksonians: The Radical Leaders of the Early Labor Movement* (Albany, 1967).

Reed, Louis, *The Labor Philosophy of Samuel Gompers* (New York, 1930).

Rees, Albert, *Real Wages in Manufacturing, 1890–1914* (Princeton, 1961).

Ringenbach, Paul, *Tramps and Reformers, 1873–1916: The Discovery of Unemployment in New York* (Westport, Conn., 1973).

Robbins, Edwin, *Railway Conductors: A Study in Organized Labor* (New York, 1914).

Robinson, Donald, *Spotlight on a Union: The Story of the United Hatters, Cap and Millinery Workers International Union* (New York, 1948).

Robinson, Jesse, *The Amalgamated Association of Iron, Steel and Tin Workers* (Baltimore, 1920).

Rosen, S. McKee, and Rosen, Laura, *Technology and Society: The Influence of Machines in the United States* (New York, 1941).

Roy, Andrew, *A History of the Coal Miners of the United States* (Columbus, 190[3]).

Salter, W. E. G., *Productivity and Technical Change* (London, 1966).

Saxton, Alexander, *The Indispensable Enemy: Labor and the Anti-Chinese Movement in California* (Berkeley, 1971).

Schluter, Hermann, *The Brewing Industry and the Brewery Workers' Movement in America* (Cincinnati, 1910).

Schmookler, Jacob, *Invention and Economic Growth* (Cambridge, 1966).

Segal, Martin, *The Rise of the United Association: National Unionism in the Pipe Trades, 1884–1924* (Cambridge, 1970).

Sennett, Richard, *Families Against the City: Middle Class Homes of Industrial Chicago, 1872–1890* (Cambridge, 1970).

Smelser, David, *Unemployment and American Trade Unions* (Baltimore, 1919).

Smelser, Neil, *Social Change in the Industrial Revolution: An Application of Theory to the British Cotton Industry* (Chicago, 1959).

Staley, Eugene, *History of the Illinois State Federation of Labor* (Chicago, 1930).

Stecker, Margaret, "The Founders, the Molders, and the Molding Machine," *Quarterly Journal of Economics*, XXXII (February 1918), 278–308.

Stockton, Frank, *The International Molders Union of North America* (Baltimore, 1921).

Strassmann, W. Paul, *Risk and Technological Innovation: American Manufacturing During the Nineteenth Century* (Ithaca, 1959).

Sullivan, William, *The Industrial Worker in Pennsylvania, 1800-1840* (Harrisburg, 1955).

Taber, Martha V., *A History of the Cutlery Industry in the Connecticut Valley*, Smith College Studies in History, Vol. 41, (Northampton, 1955).

Taft, Philip, *The A. F. L. in the Time of Gompers* (New York, 1957).

Thernstrom, Stephan, *Poverty and Progress: Social Mobility in a Nineteenth-Century City* (Cambridge, 1964).

Thompson, E. P., *The Making of the English Working Class* (London, 1963).

Tracy, George, *History of the Typographical Union* (Indianapolis, 1913).

Trattner, Walter, *Crusade for the Children: A History of the National Child Labor Committee and Child Labor Reform in America* (Chicago, 1970).

Ulman, Lloyd, *The Rise of the National Trade Union* (Cambridge, 1955).

Virtue, G. O., "The Co-operative Coopers of Minneapolis," *Quarterly Journal of Economics*, XIX (August 1905), 527-544.

Walker, Roger, "The A. F. L. and Child-Labor Legislation: An Exercise in Frustration," *Labor History*, XI (Summer 1970), 323-340.

Ware, Norman, *The Industrial Worker, 1840-1860* (Boston, 1924).

————, *The Labor Movement in the United States, 1860-1890* (New York, 1929).

Wieck, Edward, *The American Miners' Association: A Record of the Origin of Coal Miners' Unions in the United States* (New York, 1940).

Wolff, Leon, *Lockout: The Story of the Homestead Strike of 1892* (New York, 1965).

Wyllie, Irvin, *The Self-Made Man in America: The Myth of Rags to Riches* (New Brunswick, N. J., 1954).

Yearley, Clifton, Jr., *Enterprise and Anthracite: Economics and Democracy in Schuylkill County, 1820-1875* (Baltimore, 1961).

Yellowitz, Irwin, *Labor and the Progressive Movement in New York State, 1897-1916* (Ithaca, 1965).

————, "The Origins of Unemployment Reform in the United States," *Labor History*, IX (Fall 1968), 338-360.

# INDEX

American Emigrant Aid Society, 130,
    131
American Federation of Labor, 23,
    35, 44, 113-144, 116, 118, 125,
    137, 139, 161n; New York State,
    140. See also Gompers, Samuel.
Andres, Claus, 158n
Apprentices: importance of limits on,
    5, 6, 13, 14, 22, 29, 44, 68, 95-
    98, 101-102, 104, 105, 128,
    138
    local option, 96, 101
    labor, supply of, 68, 95-97, 100-
      102, 104, 114-115, 124
    overproduction, 14-15, 49-50
    reformers, 103-104
    state action, 97-98
    success of union rules, 97
    unemployment, 96-97
    wages, 95-96, 102
Atlantic Cotton Mills, 122

Baker, Herbert, 145n
Bates, John, 51
Bemis, Edward, 91
Bigham, Thomas, 41, 43, 156n
Bishop, James, 122
Blend, Frederick, 65-66
Boycotts, 101
Buchanan, Joseph, 125
Butler, Ben, 159n

Campbell, James, 100
Campbell, L. R., 121

Cameron, Andrew, 10, 11, 121
Cannon, William, 69
Carnegie, Andrew, 90, 91
Carpenters and Joiners, United Brother-
    hood of, 101, 139-140, 161n;
    Albany local, 274, 31
Chicago, Council of Trades and Labor
    Unions, 22
Child labor, 57, 63, 101, 128, 138
    effect upon workers, 105-109
    legislation, 93, 113-114
    mechanization, 72, 107-108
    reasons for, 108
    reformers, 103, 106-109, 111
    wages, 107-110
Cigar Makers' International Union, 64-
    71, 82
*Cigar Makers' Journal,* 133
Coffey, John, 138
Commons, John, 33, 35
Connolly, M. D., 102
Contract labor, see Immigration, import-
    ed workers
Cooperatives: building societies, 42
    producers, 39-40, 71, 84, 108
Coopers' International Union, 71-73,
    123

Davis, John, 48, 53
Debs, Eugene, 93, 108
Depressions, 14, 22, 25, 47, 53, 56, 57,
    68, 69, 73, 108
Devine, Edward T., 34
Division of labor: cigar makers, 65-71,
    73-74

179

*Irwin Yellowitz is Professor of History at his alma mater, the City College of the City University of New York. His Ph.D. was earned at Brown University. Having previously published numerous articles and reviews on American labor history, as well as two books on the subject:* Labor and the Progressive Movement in New York State, 1897–1916 *(1965), and* The Position of the Worker in American Society, 1865–1896 *(1969)–Dr. Yellowitz is a recognized authority in the field.*